Springer Series in Physical Environment

6

Alexander Lattermann

System-Theoretical Modelling in Surface Water Hydrology

With 83 Figures

Springer-Verlag
Berlin Heidelberg GmbH

Dr.-Ing. Alexander Lattermann
Am Fuchsloch 9
W-3260 Rinteln 5
Federal Republic of Germany

Cover Illustration: San Francisco Bay near Foster City, CA. UHAP, 1:33000, CIR (1978).
By the courtesy of H. G. Gierloff-Emden.

ISSN 0937-3047
ISBN 978-3-642-83821-7   ISBN 978-3-642-83819-4
DOI 10.1007/978-3-642-83819-4

Library of Congress Cataloging-in-Publication Data
Lattermann, Alexander, 1944 – System-theoretical modelling in surface water hydrology/
Alexander Lattermann. p. cm. – (Springer series in physical environment, ISSN
0937-3047; 6) Includes bibliographical references (p.   ) and index.
ISBN 978-3-642-83821-7

1. Hydrology – Mathematical models. I. Title. II. Series. GB656.2.M33L38 1991
551.48'01'1 – dc20   91-18951 CIP

© Springer-Verlag Berlin Heidelberg 1991
Softcover reprint of the hardcover 1st edition 1991

Typesetting: Thomson Press (India) Ltd., New Delhi
32/3145-543210 – Printed on acid-free paper

Preface

Several years ago, I was Assistant Professor of hydrology lecturing to post-graduate students at the Federal University of Paraiba, Brazil. The first course there treated the hydrology of surface water on the Earth, whereby the understanding of the natural process was emphasized. In the second course, simulation techniques of some of the hydrological quantities involved in that hydrological process were presented. The main theme proved to be that methods for hydrological modelling rely on the mathematical methods that are available and applicable.

To solve actual problems of water policy and environment the simulation technique must be improved by modelling particular units most satisfactorily adapted to the hydrological process. The success of such an improved technique depends both on more information about the phenomenon from measurements, and the availability of advanced mathematical methods.

The development in the construction of mechanical and electronic equipment and the application of computer technology result in the installation of automatic gauges with high recording precision for data requisition and subsequent data processing. This leads to more information on the behaviour of the phenomenon.

For mathematical methods, however, development is not always conformable to the equipment. Therefore, it is useful to consider the available methods and to show their possibilities in hydrological modelling, including a scientific evaluation of their potentials and limitations.

The linear system model, widely used in other disciplines such as electrical and mechanical engineering, is also a principle in surface water hydrology, and its treatment by advanced mathematical methods is brought into focus by improved measurement data.

This book is intended to give directions on dealing with elements modelling runoff processes in a river basin system. It is a textbook for courses in hydrological modelling at universities and engineering academies and, such as, it is useful not only to practitioners in water resource management, but also to professors, graduate and post-graduate students.

In the main, the book is concerned with the system-theoretical approach to modelling the rainfall-runoff process in hydrology. Firstly, the hydrological variables involved in the process are prepared for use in a linear system-theoretical model. Methods to determine the mean areal precipitation and baseflow separation are discussed.

The main part of the book deals with the convolution process in the theory of linear systems and its hydrological application in deterministic models. Through the use of both the time and frequency domains, time functions of precipitation are transformed into corresponding discharge functions.

The identification of characteristics of the watershed deterministically modelled by linear systems is based on methods of the mathematical transform technique (Laplace, Z-transformation). Stochastic hydrological processes are also used for system identification.

Additionally, two types of non-linear system models are presented to enhance the modelling of the rainfall-runoff process in surface hydrology. The usefulness and reliability of the methods are discussed in case studies.

To support the calculation of output functions of linear models, several computer programs are included in the text, and many examples and figures are given to make the models technically transparent.

It is my pleasure to express my sincere gratitude to Mrs. Maria Kaune for her encouragement and valuable comments on the first review of the text, and likewise to my son Rolf A. Lattermann for his assistance in developing computer programs. In particular, I am indebted to Professor Kurt Lecher of the Hanover University who read the preliminary version of the first chapters of the text.

Finally, I must record my appreciation of the advice and help received from many people during the preparation of this book.

Rinteln, July 1991   A. Lattermann

Contents

1 Introduction

The fact that water is an indispensable resource, essential for life on earth, makes its effective management an important goal of society.

Environmental quality depends directly on the water resources available. Improved techniques in planning, managing and developing will result only if we are able clearly to understand the physical systems of water occurrences, to find out the characteristics of which is a challenge to the hydrologist.

The problem in physics and in environmental science is to describe the occurrence in nature in terms of mathematical methods based on measurements of high accuracy.

Reliable mathematical models are necessary to evaluate and forecast the performance of hydrological systems under expected or known conditions.

The effect of water constructions such as dams, drainage systems, channelization etc. must be known before the works are carried out.

Because of the need of mathematical forecasting, in technical work mathematics and physics are inseparable.

By common consent mathematics is the most proper tool to describe processes in nature.

This basis is a justification for developing mathematical methods in physics to benefit the environment. Hydrology deals with the water of the earth, in or below the land surface and in the atmosphere.

The satisfactory allocation of water in quality and quantity are very important to municipal and industrial water supply, flood control and reservoir design.

Such utilizations of water resources are coupled with intervention in the nature of river systems. The type, mode and construction of water works depend on both the behaviour of the river system and the kind of action, if human activities and the reactions of nature are to be considered equally. The behaviour of a river system can be mainly characterized by the relationship between precipitation and runoff. Supposing that the relationship is known, i.e. a reliable mathematical model exists, this knowledge can be used to avoid damage to water resources in planning management by simulating the future situation before the water works or other interventions are carried out. The time span of prediction is different and depends on the nature of the system and the information available on its behaviour in the past. Such characteristics in time and space can be registered by physical measurements. With more measurement data, it is in general possible to make wider extrapolations and obtain better simulation results.

In like manner, if the relationship between precipitation and runoff is unknown, the quality of the mathematical model mainly depends on the measurement data by which it is calibrated.

Nevertheless, the first step is to decide which mathematical method will be suitable to solve the physical problem. For complex systems and less adequate data crudely empirical formulas may be sufficient; but with more accuracy in measurement and more information data rational equations based on physical principles should be developed.

The evaluation of the relationship between rainfall and runoff is mainly influenced by the measurement data available and the mathematical methods applied.

The qualitative consideration of the hydrological cycle shows the simple fact that runoff and evaporation in a catchment area are caused by precipitation.

Qualitatively it must be taken into account that precipitation, runoff and evaporation are time- and space-dependent.

Mathematically, such physical quantities are functions of several variables.

These physical quantities of the hydrological cycle are influenced by the behaviour of the hydrological catchment area. This influence is first considered without any interdependence of the quantities of the hydrological cycle:

The process of precipitation $P(t, x, y)$, and in the same manner the runoff process $Q(t, x, y)$ can be modelled mathematically as time- and area-dependent by use of three-dimensional functions.

Graphical illustration of these functions provides the characteristics of the process. According to the field of application, choice can be made between representing the hydrological variable as dependent on time with fixed local or space values (time functions, Euler presentation) or on local or space variability at a fixed point in time (snapshot, Lagrange presentation).

The graphical illustration of a flood wave, for example, shows plots in the form of a wave for both presentations.

From the assumption that the quantities of the hydrological cycle are inter-dependent, it follows that:

The characteristic of the catchment area providing interdependencies between precipitation and runoff can be mathematically described by two-dimensional functions.

Such a relationship can be expressed by a functional synonym as an operator (H). The operator transforms the function of precipitation into another, runoff.

This transformation is called a process, and the surroundings in which the process develops is called a system. Thus, the precipitation-runoff process takes place in the system of the catchment area, consisting of soil, soil layers and plants.

A system is always characterized by input and output functions.

The structure of the soil in the catchment area can be divided into three zones or subsystems:

– soil surface system, where the precipitation runs off to the river without penetration in the soil;
– soil subsurface system, where the precipitation and air together pass through a layer in the upper reach of the soil.
– soil groundwater system, where the water of the precipitation coming from the subsurface system and being under hydrostatic pressure, moves to the river or remains in caverns in the rock.

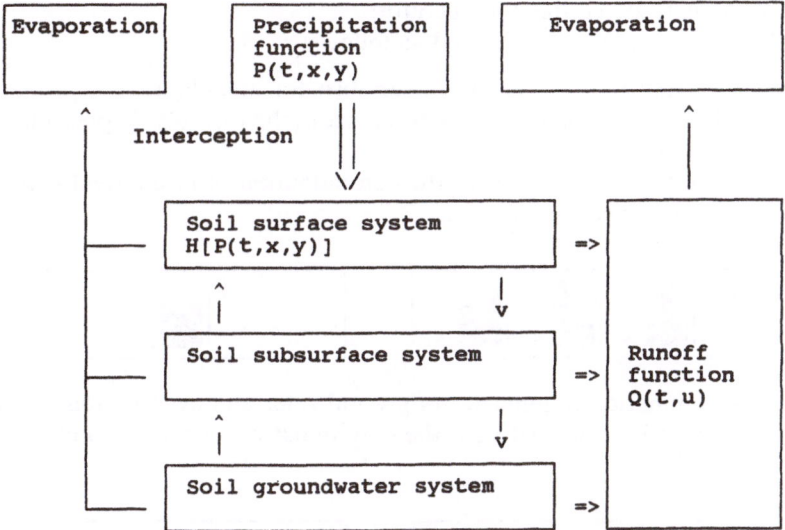

Fig. 1.1 Schematic diagram of systems in the hydrological cycle

In this text we are concerned with the mathematical modelling of the precipitation-runoff process in the soil surface system. Nevertheless, the relevant position of this process in the total soil system should also be taken into account.

This relationship is shown in the following:

– symbolic representation

or more briefly in the form of:

– mathematical representation

$$H[p(t, u)] = q(t, u).$$

One main object in this text is to determine the operator of the rainfall-runoff relationship to make the process calculable and prospectable.

The functions involved in the process are parts of the hydrological cycle and have to be specified in their effects only to the soil surface system. Furthermore, for two-dimensional functions it must be defined how the graphical illustration of the functions should be done either as Euler representation or Lagrange.

Finally, it has to be considered which mathematical theories or algorithms are available to determine the operator.

The procedure can be planned as follows:

I. Preparation of hydrological variables for modelling:
   – Elimination of the local dependence of the variables (representation as time functions, Euler).
   These models deal with the determination of mean areal precipitation in only one catchment area.

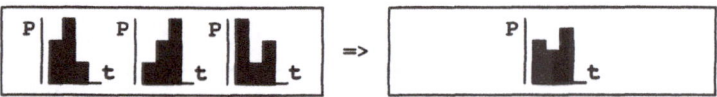

   – Elimination of influence of groundwater and evaporation on the runoff.
   These models deal with the determination of direct runoff and baseflow rates.

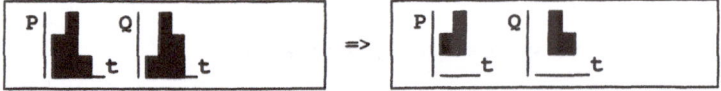

II. System-theoretical treatment of hydrological variables: Use of mathematical methods to determine the operator of the system. Measured data historically known are used to calibrate the model, i.e. the calculation of the parameter in the kernel function of the system (analysis).

With known parameters, i.e. the kernel function, actually measured data are used to simulate a future situation with the model to obtain prospectives for planned water works, etc. (synthesis)

These models deal with the transformation of precipitation in runoff.

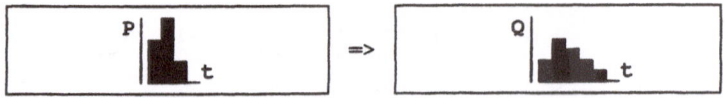

2 Preparation of Hydrological Variables for Modelling

2.1 Mean Areal Precipitation

The areal distribution of precipitation in a watershed represents in general a time- and space-dependent function.
If the time is discretized and it is regarded as only one time step, at the end of the step interval any amount of precipitable water is distributed over the watershed. It is imaginable that this precipitable water, which may be held in many slim cylinders standing close together in the catchment, forms a relief map of the precipitation. The height of the water is called rainfall depth.

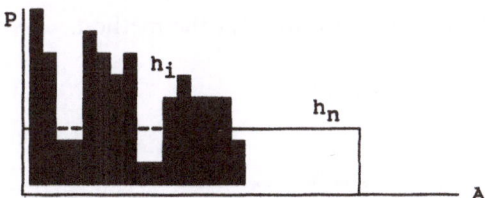

Fig. 2.1 Representation of areal rainfall distribution h_i and mean areal precipitation h_n

$h_i(x, y, t)$: Rainfall depths, depending on the position in the watershed and on time
$\quad h_n(t)$: Average depth over the watershed, depending only on time
$\qquad$ A: Catchment area (watershed)
$\qquad$ P: Precipitation.

The amount of precipitation (V) can be determined as follows:

$$V(t) = h_n(t) * A = \int\int h_i(x, y \cdot t) \, dx \, dy. \tag{2.1}$$

This relation, in which average rainfall depths and distributed rainfall depths occur, can be used to determine the average rainfall depth. Thus, this quantity depends only on time, and the space dependence of rainfall is eliminated by averaging over the area.

However, the use of this formula supposes that the area and function of the distributed rainfall are known. Unfortunately, the function of distributed rainfall is unknown. It is neither practical nor economical to install at each watershed an infinite number of rainfall gauges standing very close together to obtain measurable values of this function. Consequently, all values measured from raingauge stations, depending on the network density, can be used as only an approximation of this function. Each measured rainfall depth (h_i) of a single raingauge station is assumed to be representative for rainfall in the surrounding area (A_i). The extension of this area depends on the density of the raingauge network.

Thus, the average rainfall depth (mean areal precipitation) can commonly be determined by the formula:

$$h_n(t) = [\Sigma h_i(t) A_i]/A. \tag{2.2}$$

The following different methods show how representative the areal rainfall is assumed to be over a particular area surrounding a raingauge station, and what extension of the area is attached to it.

Isohyetal Method

Isohyetal maps are prepared for a catchment area by interpolation of raingauges. Weighted averages of depths between the isohyets are used to estimate areal precipitation.

Example demonstrating the method:

Given problem:

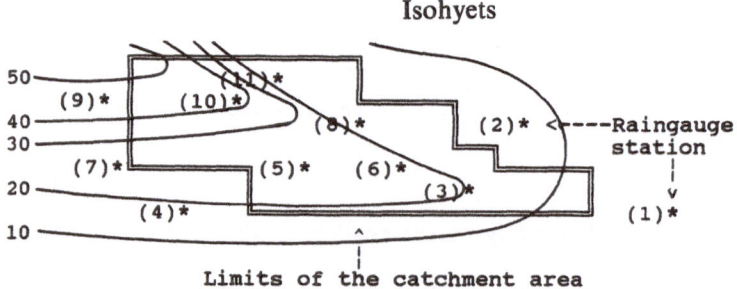

Fig. 2.2 Boundary of the catchment area

Values of measurements:

   Catchment area: $A = 626\,UA$ (units of area)

Rainfall depths measured at station in one time interval of e.g. 1 day.

Station $(1) = 6.5\,\text{mm}, (2) = 14.6\,\text{mm}, (3) = 19.2\,\text{mm}, (4) = 15.4\,\text{mm}$

$(5) = 26.9\,\text{mm}, (6) = 28.9\,\text{mm}, (7) = 28.2\,\text{mm}, (8) = 17.5\,\text{mm}$

$(9) = 50.0\,\text{mm}, (10) = 45.0\,\text{mm}, (11) = 19.5\,\text{mm}.$

Linear distance between the raingauge stations:

Station $(1) - (2) = 2.0\,\text{UL}$ (units of length)

$(1) - (3) = 2.0$ $-"-$

$(1) - (4) = 4.5$ $-"-$

. . . .

. . . .

Desired: Areal precipitation $h_n(t)$

Solution:

1. Calculation of isohyets (lines of equal rainfall depth): The position of the 10-mm isohyet is calculated by the following interpolation equation:

$$P_i/D_i = P_{10}/D_x$$

$P_i =$:Difference of rainfall depths between considered gauge stations

$D_i =$:Distance of length between considered gauge stations

$P_{10} =$:Difference of rainfall depths between 10-mm value and the value measured at the considered station.

$D_x =$:Desired distance of length between the considered station and the 10-mm isohyet.

$(1) - (2): D_x = 2.0(10.0 - 6.5)/(14.6 - 6.5) = 0.86\,\text{UL}$

$(1) - (3): D_x = 2.0(10.0 - 6.5)/(19.2 - 6.5) = 0.55\,\text{UL}$

$(1) - (4): D_x = 4.5(10.0 - 6.5)/(15.4 - 6.5) = 1.77\,\text{UL}.$

With station (1) as a starting point, the line of the 10-mm isohyet is reached in 0.86 length units in the direction of station (2), in 0.55 length units in the direction of station (3), and in 1.77 length units in the direction of station (4).

This calculation has to be made for all stations and at reasonable isohyet distances to find the points which form the isohyets by drawing a curve.

2. Calculation of partial areas (A_i) between the isohyets: The determination can be made graphically by use of a planimeter.

For this example the following values are obtained:

$A_1 < 10\ ;\ A_1 = 31\,\text{UA}$

$10 < A_2 < 20\ ;\ A_2 = 193\,\text{UA}$

$20 < A_3 < 30\ ;\ A_3 = 196\,\text{UA}$

$30 < A_4 < 40\ ;\ A_4 = 116\,\text{UA}$

$40 < A_5 < 50\ ;\ A_5 = 77\,\text{UA}$

$50 < A_6 \qquad ;\ A_6 = 13\,\text{UA}.$

3. Calculation of the areal precipitation $h_n(t)$:

Using the formula: $h_n(t) = 1/A(\Sigma h_i A_i)$ we obtain:

$h_1 = (6.5 + 10.0)/2 = 8; h_2 = (10.0 + 20)/2 = 15$

. . . . , ...

$h_6 = (50 + \cdots)/2$

h_i mm	A_i (UA)	$h_i A_i$ mm (UA)
8	31	248
15	193	2895
25	196	4900
35	116	4060
45	77	3465
53	13	689
	626	$16217 = \Sigma h_i A_i$

$h_n(t) = 16217/626 = 25.90 \, mm \approx 26 \, mm.$

The calculation must be repeated for each considered time interval to obtain the time distribution of areal precipitation.

Advantages of the method:
This method is the most accurate approach for determining average precipitation over an area. The distance between the isohyets can be made arbitrarily narrow, so that no marked steps occur in the relief of the rainfall distribution. The weights for averaging are determined anew for each time interval.
Disadvantages of the method:
The method requires expensive calculations because the isohyetal map differs for each time interval considered. To design isohyets, topographical factors need careful attention.
Aid to calculation:
The calculation can be made more easily by using the computer program PMSB1 (Appendix 1).

Thiessen Method
By this method the catchment area is divided into polygonal subareas with raingauge stations as centres. The subareas represent weights, and the mean areal precipitation is estimated by a weighted mean.

Example for demonstration of the method:
Given problem:

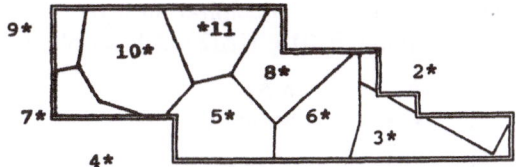

Fig. 2.3 Boundary of the catchment area

Values of measurement:

Catchment area = 626 AU

Rainfall depths measured on station in one time interval, e.g. 1 day.

Station (1) = 6.5 mm, (2) = 14.6 mm, (3) = 19.2 mm, (4) = 15.4 mm

(5) = 26.9 mm, (6) = 28.9 mm, (7) = 28.2 mm, (8) = 17.5 mm

(9) = 50.0 mm, (10) = 45.0 mm, (11) = 19.5 mm.

Linear distance between the raingauge stations:

Station (1) − (2) = 2.0 UL

(1) − (3) = 2.0 UL

(1) − (4) = 4.5 UL

. . . .

Desired: Areal precipitation $h_n(t)$

Solution:

1. Calculation of polygonal subareas: Each raingauge location in the map is connected by straight lines, and on half these lines perpendiculars are raised forming sectors with the raingauges as their centres.

   Each area of the sectors is associated with a station and used as a weight in estimating average depths. The subareas can be determined in practice by use of a planimeter or by a computer program, if the catchment area is available in a digitized form.

   In the example the following values are obtained:

Station —	Polygonal subareas UA
1	7
2	20
3	90
4	—
5	109
6	82
7	40
8	100
9	42
10	76
11	60

2. Calculation of the areal precipitation $h_n(t)$:
   Using the formula: $h_n(t) = 1/A(\Sigma h_i A_i)$, we obtain:

h_i mm	A_i UA	$h_i A_i$ mm UA
6.5	7	45.0
14.6	20	292.0
19.2	90	1728.0
15.4	—	—
26.9	109	2932.1
28.9	82	2369.8
28.2	40	1128.0
17.5	100	1750.0
50.0	42	2100.0
45.0	76	3420.0
19.5	60	1170.0
	626	$16935.4 = \Sigma h_i A_i$

$h_n(t) = 16935.4/626 = 27.05 \text{ mm} \approx 27 \text{ mm}.$

The calculation must be repeated for each considered time interval to obtain the time distribution of areal precipitation.

Advantages of the method:

A different distribution or raingauge locations in the watershed can be taken into account by the influence of weights.

The network of polygonal subareas is fixed for a given gauge configuration in each time interval.

Disadvantages of the method:

The weights are uniformly distributed, and where the limits of the subareas meet discontinuities occur, artificially caused by the method. The nature of rainfall is a continuous process.

Aid to calculation:

The calculation can be made more easily by using the computer program PMSB2 (Appendix 2).

Method of the Arithmetic Average

In the case of uniformly distributed raingauge locations in the watershed, or provided that all measurements of the stations have the same weighted influence for averaging, all subareas of the Thiessen method are of equal size. The calculation of areal precipitation can then be simplified in the following way:

$$A_1 = A_2 = \cdots = A_i = A_n \tag{2.3}$$

$$A_1 + A_2 + \cdots + A_i + A_n = nA_i = A \tag{2.4}$$

$$A_i/A = A_i/nA_i = 1/n \tag{2.5}$$

$$h_n = 1/n(\Sigma h_i). \tag{2.6}$$

The result is the formula of arithmetic average.
Example to demonstrate the method:
Given problem:

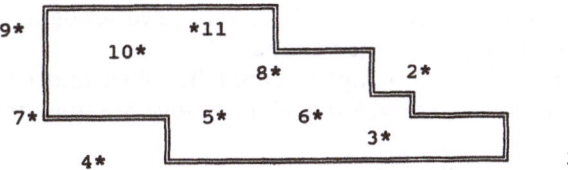

Fig. 2.4 Boundary of the catchment area

Values of measurements:
Catchment area: $A = 626 \, \text{UA}$
Rainfall depths measured on station in one time interval, e.g. 1 day.

Station $(1) = 6.5 \, \text{mm}, (2) = 14.6 \, \text{mm}, (3) = 19.2 \, \text{mm}, (4) = 15.4 \, \text{mm}$
$(5) = 26.9 \, \text{mm}, (6) = 28.9 \, \text{mm}, (7) = 28.2 \, \text{mm}, (8) = 17.5 \, \text{mm}$
$(9) = 50.0 \, \text{mm}, (10) = 45.0 \, \text{mm}, (11) = 19.5 \, \text{mm}.$

Desired: Areal precipitation $h_n(t)$
Solution:
Calculation of the areal precipitation:
Using the formula:

$$h_n(t) = 1/n(\Sigma h_i),$$ (2.7)

we obtain:

Station	h_i	n
—	mm	—
1	6.5	—
2	14.6	—
3	19.2	1
4	28.2	—
5	26.9	2
6	28.9	3
7	28.2	—
8	17.5	4
9	50.0	—
10	45.0	5
11	19.5	6

$\Sigma 271.7$

$h_n(t) = 271.7/11 = 19.79 \, \text{mm}.$

The calculation must be repeated for each considered time interval to obtain the time distribution of areal precipitation.

Advantages of the method:

The method is very simple and the results are satisfactory, if the raingauges are uniformly distributed and the watershed topography is flat.

Disadvantages of the method:

Influences of orography cannot be taken into consideration. For differently distributed and widespread raingauge stations, this method gives unreliable results.

Aid to calculation:

The calculation can be made more easily by using the computer program PMSB3 (Appendix 3).

Grid Point Method

This method differs from the previous methods as follows:

In a weighted mean for calculating the areal precipitation no weights are used that correspond to a subarea of the raingauge locations. Rainfall weights, not areal weights are used for calculating.

The catchment area is divided by a square grid of a determined size.

Each square is associated with the weights, depending on the rainfall depths and the reciprocal distance of the proximate gauge station. The arithmetical mean of the associated values of squares represents the areal precipitation.

The procedure is mathematically formulated as follows:

$h_n(t) = 1/n[\Sigma h_j(t)]$ Arithmetical mean of the weights, h_j, associated with the squares

$h_j(t) = \sum_{i=1}^{i=4} v_{ij} h_i(t)$ Influence of rainfall depth, v, of the four proximate gauge stations on the square, partial weight 1

$v_{ij} = (1/d_i^2) \Big/ \sum_{i=1}^{i=4} (1/d_i^2)$ Influence of distance, d, of the four proximate gauge stations on the square, partial weight 2

Example to demonstrate the method:

Given problem:

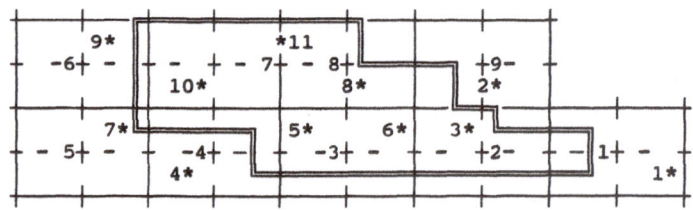

Fig. 2.5 Boundary of the catchment area

Values of measurements:

Catchment area: 626 UA (units of area)

Rainfall depths measured on station in one time interval e.g. 1 day.

Station $(1) = 6.5$ mm, $(2) = 14.6$ mm, $(3) = 19.2$ mm, $(4) = 15.4$ mm

$\qquad (5) = 26.9$ mm, $(6) = 28.9$ mm, $(7) = 28.2$ mm, $(8) = 17.5$ mm

$\qquad (9) = 50.0$ mm, $(10) = 45.0$ mm, $(11) = 19.5$ mm.

Distances between the squares and the proximate station in dependence on four quadrants following from a system of coordinates in the centre of the squares:

Square —	Quadrant dI cm	Quadrant dII cm	Quadrant dIII cm	Quadrant dIV cm
1	—	2.7	8.3	1.0
2	1.4	0.5	5.7	3.5
3	1.0	0.8	3.0	6.2
4	1.7	1.8	0.7	8.5
5	0.9	—	—	2.0
6	0.8	—	—	1.5
7	1.4	2.3	0.5	2.1
8	—	1.3	1.5	0.5
9	—	3.8	2.3	0.4.

Desired: Areal precipitation $h_n(t)$

Solution:

1. Determination of the influence exerted on squares by distances of the stations:

$$v_{ij} = (1/d_i^2) \Big/ \sum_{i=1}^{i=4} (1/d_i^2). \qquad (2.8)$$

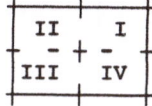

Each square consists of four quadrants.

d_i = Distance between the centre of the square and the proximate station in quadrant i (i = 1–4).

v_{ij} = Weight of distance influence of square j in quadrant i

For square 1, for example, the following values are obtained:

$\qquad$ dI $= -$; dII $= 2.7$ cm; dIII $= 8.3$ cm; dIV $= 1.0$ cm:

$\qquad$ vI1 $= -$

$\qquad$ vII1 $= (1/2.7^2)/\Sigma(1/2.7^2 + 1/8.3^2 + 1/1.0^2) = 0.12$

$\qquad$ vIII1 $= (1/8.3^2)/\Sigma(1/2.7^2 + 1/8.3^2 + 1/1.0^2) = 0.01$

$\qquad$ vIV1 $= (1/1.0^2)/\Sigma(1/2.7^2 + 1/8.3^2 + 1/1.0^2) = 0.87$

Square	Quadrant I			Quadrant II			Quadrant III			Quadrant IV		
—	St.	dI	vI_j	St.	dII	vII_j	St.	dIII	$vIII_j$	St.	dIV	vIV_j
j	—	cm	—	—	cm	—	—	cm	—	—	cm	—
1	—	—	—	2	2.7	0.12		8.3	0.01		1.0	0.87
2		1.4			0.5			5.7			3.5	
3		1.0			0.8			3.0			6.2	
4		1.7			1.8			0.7			8.5	
5		0.9			—			—			2.0	
6		0.8			—			—			1.5	
7		1.4			2.3			0.5			2.1	
8		—			1.3			1.5			0.5	
9		—			3.8			2.3			0.4	

2. Determination of the influence exerted on squares by rainfall depths of the stations:

$$h_j = \sum_{i=1}^{i=4} v_{ij} h_i \qquad (2.9)$$

h_i = Rainfall depths in the quadrant i of proximate station
h_j = Weight of rainfall influence in square j

Example:
For square 1, the following value is obtained:

$$h_1 = \Sigma(0.12 \cdot 14.6 + 0.01 \cdot 15.4 + 0.87 \cdot 6.5) = 7.56 \, mm$$

Square	Quadrant		Quadrant		Quadrant		Quadrant		
—	h_i	vI_j	h_i	vII_j	h_i	$vIII_j$	h_i	vIV_j	Σ
j	mm	—	mm	—	mm	—	mm	—	
1	—	—	14.6	0.12	15.4	0.01	6.5	0.87	7.56
2									
3									
4									
5									
6									
7									
8									
9									

Σ 223.19

3. Calculation of the areal precipitation $h_n(t)$:

$$h_n(t) = 1/m[\Sigma h_j(t)] \qquad (2.10)$$

m = Number of total squares
h_j = Rainfall weight in square j

For this example we obtain:

$h_n(t) = 1/9 \cdot 223.19 = 24.8$ mm.

The calculation must be repeated for each considered time interval to obtain
a time distribution of areal precipitation.
Advantages of the method:
The size of the quadratic grid can arbitrarily be made narrow so that the
transitions are smoothed at the limits of the sphere of influence of the raingauge
stations. These conform to natural rainfall development.
Disadvantages of the method:
With growing narrowness of the grid, the extent of the calculations increases.
Aid to calculation:
The calculation can be made more easily by using the computer program
PMSB4 (Appendix 4).

Kriging Method

To determine the mean areal precipitation let us basically use the well-known
formula of average rainfall depths:

$$h_n(t) = [\Sigma h_i(t)A_i]/A. \tag{2.11}$$

In this formula the quotients of $A_i/A = \Phi_i$ represent weights, and in consequence
the mean areal precipitation is a weighted mean with the constraint that $\Sigma\Phi_i = 1$.
Considering a constant time interval at a fixed time we can write:

$$h_n(t) = h_n = \Sigma\Phi_i h_i \quad \text{and} \quad \Sigma\Phi_i = 1. \tag{2.12}$$

As long as the true weights for determining the mean areal precipitation (h_n)
are unknown, we should speak of estimates for (h_n) and denote it by (h'_n).
To estimate (h_n), there will then be considered a weighted mean of the n available
or known values of precipitation depths at the gauge stations in a certain time
interval.

$$h'_n = \Sigma\Phi_i h_i \tag{2.13}$$

The problem is to find out the weights so that the best estimation is possible
in the sense that the error between h'_n and h_n is zero.
The rainfall depths h_i measured at a raingauge station can be interpreted as a
sample value of a random variable of a stationary stochastic process of rainfall
events.
A stochastic process $h(x, y, t, z)$ can generally be defined as a family of real
functions, where t, x, y, are real arguments (t = time; x, y = coordinates of a point
in an area) and z stands for any measurable state. This process asserts different
facts, depending on the behaviour of its variables. If the variable z is fixed, the
process consists of a real function $h(x, y, t)$, and if the variables x, y, t are fixed,

the process represents a random variable. A random variable is characterized by a probability density function with statistical parameters like the expected value (E), the variance (σ^2), etc.

A stochastic process is said to be local stationary if its joint probability density function (PDF) at arbitrary points (x_i, y_i) is invariant for translation at all the points in the area under study.

Considering this invariance for translation only for the first two moments of PDF, we can then speak of a weak stationary process. The expected value is constant:

$$E(h_i) = h = const. \tag{2.14}$$

The covariance between the two points h_i and h_j in the considered area is independent of the position of h_i and h_j, but depends only on the distances $(x_i - x_j, y_i - y_j)$ between them:

$$Cov(h_i, h_j) = E[(h_i - h)(h_j - h)] = f(x_i - x_j, y_i - y_j). \tag{2.15}$$

A stochastic process where the condition for weak stationarity is related only to distances $(x_i - y_j, y_i - y_j)$, but not to the sample values, is of a weaker stationary, and is called a stochastic process of "intrinsic hypothesis".

It can be expressed by writing:

$$E[h(x_i - x_j, y_i - y_j)] = 0 = const. \tag{2.16}$$

$$Cov[h(x_i - x_j, y_i - y_j)] = E\{[(h_i - h_j) - 0][(h_i - h_j) - 0]\} \tag{2.17}$$

$$= E[(h_i - h_j)^2] \tag{2.18}$$

$$= var[h(x_i - x_j, y_i - y_j)]. \tag{2.19}$$

The variances of the sample values depending on distances represent a function that is called the variogram $2\Gamma(\tau)$, where $\tau = (x_i - x_j, y_i - y_j)$ is a vector.

$$2\Gamma(\tau) = var[h(x_i - x_j, y_i - y_j)] = E[(h_i - h_j)^2]. \tag{2.20}$$

The expression of the semi-variogram is commonly used:

$$\Gamma(\tau) = [E(h_i - h_j)^2]/2 \tag{2.21}$$

$$= (1/2n) \sum_{i=1}^{i=n} [h(i + \tau) - h(i)]^2. \tag{2.22}$$

This formula can be used to evaluate experimentally the variogram from (n) sample values of a single available realization.

If the phenomenon under study shows an isotropic behaviour the variogram is independent of the directions of the vector τ, and the function $\Gamma(\tau)$ depends only on the modulus $|\tau|$.

In spite of the anisotropic behaviour of rainfall, for the sake of simplicity we shall suppose it to be isotropic, and additionally, the rainfall process is assumed to be of a weak stationarity.

Assuming the intrinsic hypothesis for rainfall measurements, the weights for estimating the mean areal precipitation must be such that the h'_n is not over- or under-estimated (unbiased) and the mean squares error of the estimation variance is minimum. This can mathematically be formulated:

$$E(h'_n - h_n) = 0 \qquad \text{(unbiased condition)}$$
$$\text{var.}(h'_n - h_n) = \min \qquad \text{(estimation variance condition).}$$

If the constraint $\Sigma\Phi_i = 1$ for the weights holds, the estimation variance can be written as a linear combination of increments:

$$h'_n - h_n = \Sigma\Phi_i h_i - 1 h_n = \Sigma\Phi_i h_i - \Sigma\Phi_i h_n \qquad (2.23)$$

$$= \Sigma\Phi_i(h_i - h_n). \qquad (2.24)$$

The estimation variance can be written as follows:

$$\text{var.}(h'_n - h_n) = \text{var.}(\Sigma\Phi_i h_i - h_n) \qquad (2.25)$$

$$= E\{[\Sigma\Phi_i(h_i - h_n) - 0][\Sigma\Phi_j(h_j - h_n) - 0]\} \qquad (2.26)$$

$$= E[(\Sigma\Phi_i h_i - h_n)(\Sigma\Phi_j h_j - h_n)] \qquad (2.27)$$

$$= E(\Sigma\Sigma\Phi_i\Phi_j h_i h_j - 2h_n\Sigma\Phi_j h_j + h_n^2). \qquad (2.28)$$

By use of the following identity:

$$\Sigma\Phi_j(h_n - h_j)^2 = \Sigma\Phi_j h_n^2 - 2\Sigma\Phi_j h_n h_j + \Sigma\Phi_j h_j^2 \qquad (2.29)$$

$$= h_n^2 - 2h_n\Sigma\Phi_j h_j + h_j^2$$

$$\Sigma\Phi_j(h_n - h_j)^2 - h_j^2 = h_n^2 - 2h_n\Sigma\Phi_j h_j \qquad (2.30)$$

and furthermore:

$$\Sigma\Sigma\Phi_i\Phi_j(h_i - h_j)^2 = \Sigma\Sigma\Phi_i\Phi_j h_i^2 - 2\Sigma\Sigma\Phi_i\Phi_j h_i h_j + \Sigma\Sigma\Phi_i\Phi_j h_j^2 \qquad (2.31)$$

$$= h_i^2 - 2\Sigma\Sigma\Phi_i\Phi_j h_i h_j + h_j^2$$

$$\Sigma\Sigma\Phi_i\Phi_j h_i h_j = -0.5\Sigma\Sigma\Phi_i\Phi_j(h_i - h_j)^2 + 0.5h_i^2 + 0.5h_j^2 \qquad (2.32)$$

we obtain by substitution:

$$\text{var.}(h'_n - h_n) = E[-0.5\Sigma\Sigma\Phi_i\Phi_j(h_i - h_j)^2 + 0.5h_i^2 + 0.5h_j^2$$
$$+ \Sigma\Phi_j(h_n - h_j)^2 - h_j^2] \qquad (2.33)$$

$$= -0.5\Sigma\Sigma\Phi_i\Phi_j E[(h_i - h_j)^2] + \Sigma\Phi_j E[(h_n - h_j)^2]$$
$$+ 0.5E(h_i^2 - h_j^2). \qquad (2.34)$$

Finally, the estimation variance is expressed as a function of the semi-variogram:

$$\text{var.}(h'_n - h_n) = -\Sigma\Sigma\Phi_i\Phi_j\Gamma_{(i,j)} + 2\Sigma\Phi_j\Gamma_{(i,n)} + 2\Theta$$
$$\text{with} \quad \Theta = -0.5E(h_i^2 - h_j^2). \qquad (2.35)$$

This equation for the variance will now be minimized by the mean squares estimation method:

$$\delta \text{ var.}(h'_n - h_n)/\delta \Phi_j = \Gamma_{(i,n)} - \Sigma \Phi_i \Gamma_{(i,j)} + \Theta = 0 \tag{2.36}$$

with the constraint: $\Sigma \Phi_i = 1$.

This requirement for minimal estimation variance leads to a system of $(n + 1)$ linear equations to determine the $(n + 1)$ unknown quantities, n weights Φ_i and the Lagrange multiplier Θ.

In the earth sciences this method of estimation is known as kriging, after D.G. Krige, who devised it empirically for use in the South African goldfields.

The method is a means of weighted local averaging. The weights are chosen such that there are unbiased estimates with minimized estimation variance.

In this method the available measurements of a property at a number of places within a region are used to estimate the property of any one place. If this place is of the same size and shape as those where the measurements are made, the procedure for estimation is called "punctual kriging", and if the place is a larger area or block, the method is called "block-kriging".

In the case of estimating the mean areal precipitation we are dealing with "block-kriging". The area of interest or water-shed is regarded as a block. It can be covered by an artificial meshwork, and the "punctual kriging" is applied to estimate mean values at all gridpoints.

Depending on the meshwork density, the estimated values at the gridpoints situated in the centres of meshes can be assumed to be constant over the particular mesh areas.

Averaging the kriged estimates over all the meshes results in the mean areal precipitation.

The system of linear equations is best represented in matrix form:

$$\Phi_1 \Gamma(1, 1) + \Phi_2 \Gamma(1, 2) + \cdots + \Phi_n \Gamma(1, n) + \Theta = \Gamma(1, n)$$
$$\Phi_1 \Gamma(2, 1) + \Phi_2 \Gamma(2, 2) + \cdots + \Phi_n \Gamma(2, n) + \Theta = \Gamma(2, n)$$
$$\Phi_1 \Gamma(3, 1) + \Phi_2 \Gamma(3, 2) + \cdots + \Phi_n \Gamma(3, n) + \Theta = \Gamma(3, n)$$
$$\Phi_1 \Gamma(4, 1) + \Phi_2 \Gamma(4, 2) + \cdots + \Phi_n \Gamma(4, n) + \Theta = \Gamma(4, n)$$

$$\Phi_1 \Gamma(n, 1) + \Phi_2 \Gamma(n, 2) + \cdots + \Phi_n \Gamma(n, n) + \Theta = \Gamma(n, n)$$
$$\Phi_1 \qquad + \Phi_2 \qquad + \cdots + \Phi_n \qquad + 0 = \quad 1.$$
$$\bar{\bar{\Gamma}} * \vec{\Phi} = \vec{N}, \tag{2.37}$$

where:

$$\left.\begin{array}{l}\Gamma(1,1)+\Gamma(1,2)+\cdots+\Gamma(1,n)+1\\ \Gamma(2,1)+\Gamma(2,2)+\cdots+\Gamma(2,n)+1\\ \Gamma(3,1)+\Gamma(3,2)+\cdots+\Gamma(3,n)+1\\ \Gamma(4,1)+\Gamma(4,2)+\cdots+\Gamma(4,n)+1\\ \quad\cdot\cdot\qquad\cdot\cdot\qquad\cdots\qquad\cdot\cdot\quad+\cdot\\ \quad\cdot\cdot\qquad\cdot\cdot\qquad\cdots\qquad\cdot\cdot\quad+\cdot\\ \Gamma(n,1)+\Gamma(n,2)+\cdots+\Gamma(n,n)+1\\ \quad 1\quad+\quad 1\quad+\cdots+\quad 1\quad+0\end{array}\right\}=\bar{\Gamma}$$

$$\begin{array}{l}\Phi_1\\ \Phi_2\\ \Phi_3\\ \Phi_4\\ \cdot\\ \cdot\\ \Phi_n\\ \Theta\end{array}=\overset{>}{\Phi}\qquad\begin{array}{l}\Gamma(1,n)\\ \Gamma(2,n)\\ \Gamma(3,n)\\ \Gamma(4,n)=\overset{>}{N}\\ \cdot\\ \cdot\\ \Gamma(5,n)\\ 1.\end{array}$$

The matrix equation can be solved for Φ as follows:

$$\overset{>}{\Phi}=(\bar{\Gamma})^{-1}*\overset{>}{N}. \tag{2.38}$$

To demonstrate the method we should consider the following example:
Given problem:

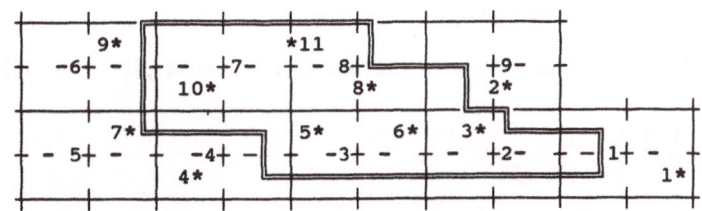

Fig. 2.6 Boundary of the catchment area

Values of measurement:
Catchment area = 626 AU
Rainfall depths measured at the station in a time interval e.g. 1 day.

Station $(1)=\;6.5\,\text{mm}, (2)=14.6\,\text{mm}, (3)=19.2\,\text{mm}, (4)=15.4\,\text{mm}$

$\quad\quad\quad (5)=26.9\,\text{mm}, (6)=28.9\,\text{mm}, (7)=28.2\,\text{mm}, (8)=17.5\,\text{mm}$

$\quad\quad\quad (9)=50.0\,\text{mm}, (10)=45.0\,\text{mm}, (11)=19.5\,\text{mm}.$

Linear distances between the raingauge stations in UL:

$\quad (1)-(2)=2.95\quad (2)-(3)=0.89\quad (3)-(4)=4.40\quad (4)-(5)=2.08$

$\quad (1)-(3)=3.05\quad (2)-(4)=5.00\quad (3)-(5)=2.40\quad (4)-(6)=3.45$

$(1) - (4) = 7.40 \quad (2) - (5) = 2.90 \quad (3) - (6) = 1.00 \quad (4) - (7) = 1.16$

$(1) - (5) = 5.43 \quad (2) - (6) = 1.61 \quad (3) - (7) = 5.40 \quad (4) - (8) = 3.13$

$$\vdots \qquad\qquad \vdots \qquad\qquad \vdots \qquad\qquad \vdots$$

$(5) - (6) = 1.40 \quad (6) - (7) = 4.40 \quad (7) - (8) = 3.88 \quad (8) - (9) \ \ = 4.04$

$(5) - (7) = 3.00 \quad (6) - (8) = 1.00 \quad (7) - (9) = 1.41 \quad (8) - (10) = 2.60$

$$\vdots \qquad\qquad \vdots \qquad\qquad \vdots \qquad\qquad \vdots$$

$(9) - (10) = 1.56 \ (10) - (11) = 1.56$

$(9) - (11) = 2.80.$

Desired : Mean areal precipitation $h_n(t)$

Solution:

1. Estimation of the semi-variogram:

$$\Gamma(\tau) = (1/2n) \sum_{i=1}^{i=n} [h(i+\tau) - h(i)]^2 \qquad\qquad (2.39)$$

$\tau = 1.8 \ UL:$ (class widths)

$\Gamma(0.9) = 102.18 \ mm^2$

$\Gamma(2.7) = 125.80 \ mm^2$

$\Gamma(4.5) = 205.20 \ mm^2$

$\Gamma(6.3) = 250.40 \ mm^2$

$\Gamma(8.1) = 490.57 \ mm^2.$

These estimated semi-variogram values can be plotted as follows:

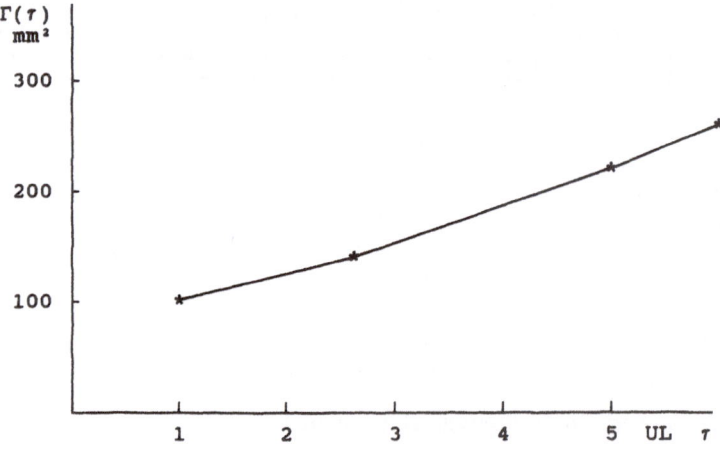

Fig. 2.7 Semi-variogram

The kriging system of linear equations requires a known semi-variogram, i.e. knowledge of the spatial variability of rainfall. The sample semi-variogram consists of discrete values, whereas only a continuous semi-variogram function corresponds with the behaviour of rainfall, that varies continuously in space. Therefore, the sample semi-variogram is used for selecting a theoretical semi-variogram model. The principal feature of semi-variograms of precipitation can be represented by linear functions of positive slopes (a) and positive constants (b) (nugget variance).

However, at the origin, the semi-variogram is zero, and for mathematical formulation it will be taken into account by the use of the Dirac function (δ) as follows:

$$\Gamma(\tau) = a(1 - \delta) + b\tau. \tag{2.40}$$

Using the regression technique for determination of the parameters (a) and (b), we can fit the theoretical model on the sample values.

τ	0.9	2.7	4.5	6.3	8.1
$\Gamma(\tau)$	102	125	205	250	490

$$E(\Gamma) = (\Sigma \Gamma)/5 = 234.8 \qquad E(\Gamma^2) = (\Sigma \Gamma^2)/5 = 74349.0 \tag{2.41}$$

$$E(\tau) = (\Sigma \tau)/5 = 4.5 \qquad E(\tau^2) = (\Sigma \tau^2)/5 = 26.4 \tag{2.42}$$

$$E(\Gamma\tau) = (\Sigma \Gamma\tau)/5 = 1372.7 \tag{2.43}$$

$$S(\Gamma) = \sqrt{E(\Gamma^2) - E^2(\Gamma)} = 138.6; \quad a = S(\Gamma\tau)/S^2(\tau) \tag{2.44}$$

$$S(\tau) = \sqrt{E(\tau^2) - E^2(\tau)} = 2.5; \qquad a = 323/6.3 = 50.4 \tag{2.45}$$

$$S(\Gamma\tau) = E(\Gamma\tau) - E(\Gamma)E(\tau) = 323.0; \quad b = E(\Gamma) - aE(\tau) = 9.5 \tag{2.46}$$

2. Setting up the kriging system of linear equations:
   The semi-variance between the gauge stations:

$\Gamma(1, \ 1) = 0$ $\Gamma(2, \ 2) = 0$
$\Gamma(1, \ 2) = 158$ $\Gamma(2, \ 3) = \ 54$
$\Gamma(1, \ 3) = 163$ $\Gamma(2, \ 4) = 261$
$\Gamma(1, \ 4) = 382$ $\Gamma(2, \ 5) = 156$

    .   .   .   .

$\Gamma(3, \ 3) = 0$ $\Gamma(4, \ 4) = 0$
$\Gamma(3, \ 4) = 233$ $\Gamma(4, \ 5) = 114$
$\Gamma(3, \ 5) = 130$ $\Gamma(4, \ 6) = 183$
$\Gamma(3, \ 6) = \ 59$ $\Gamma(4, \ 7) = \ 68$

    .   .   .   .

$\Gamma(5, \ 5) = 0$ $\Gamma(6, \ 6) = 0$

$\Gamma(5, 6) = 80$ $\Gamma(6, 7) = 231$
$\Gamma(5, 7) = 160$ $\Gamma(6, 8) = 59$
$\Gamma(5, 8) = 66$ $\Gamma(6, 9) = 251$

$\Gamma(7, 7) = 0$ $\Gamma(8, 8) = 0$
$\Gamma(7, 8) = 205$ $\Gamma(8, 9) = 213$
$\Gamma(7, 9) = 80$ $\Gamma(8, 10) = 140$

$\Gamma(9, 9) = 0$ $\Gamma(10, 10) = 0$
$\Gamma(9, 10) = 88$ $\Gamma(10, 11) = 88$

$\Gamma(11, 11) = \Gamma(0) = 0.$

The average semi-variance within the block:

$\Gamma(1, 1) = 52$ $\Gamma(2, 1) = 115$
$\Gamma(1, 2) = 151$ $\Gamma(2, 2) = 65$
$\Gamma(1, 3) = 251$ $\Gamma(2, 3) = 133$

$\Gamma(1, n) = 279$ $\Gamma(2, n) = 173$

$\Gamma(3, 1) = 171$ $\Gamma(4, 1) = 342$
$\Gamma(3, 2) = 27$ $\Gamma(4, 2) = 241$
$\Gamma(3, 3) = 101$ $\Gamma(4, 3) = 141$

$\Gamma(3, n) = 160$ $\Gamma(4, n) = 164$

$\Gamma(5, 1) = 241$ $\Gamma(6, 1) = 171$
$\Gamma(5, 2) = 141$ $\Gamma(6, 2) = 71$
$\Gamma(5, 3) = 43$ $\Gamma(6, 3) = 52$

$\Gamma(5, n) = 132$ $\Gamma(6, n) = 141$

$\Gamma(7, 1) = 392$ $\Gamma(8, 1) = 208$
$\Gamma(7, 2) = 292$ $\Gamma(8, 2) = 117$
$\Gamma(7, 3) = 191$ $\Gamma(8, 3) = 65$

$\Gamma(7, n) = 184$ $\Gamma(8, n) = 135$

$\Gamma(9, 1) = 411$ $\Gamma(10, 1) = 335$
$\Gamma(9, 2) = 314$ $\Gamma(10, 2) = 236$

$$\Gamma(9,3) = 219 \qquad \Gamma(10,3) = 140$$

$$\vdots \qquad \vdots \qquad\qquad \vdots \qquad \vdots$$

$$\Gamma(9,n) = 201 \qquad \Gamma(10,n) = 152$$

$$\Gamma(11,1) = 275$$
$$\Gamma(11,2) = 183$$
$$\Gamma(11,3) = 108$$

$$\vdots \qquad \vdots$$

$$\Gamma(11,n) = 145.$$

The kriging system of linear equations:

000	158	163	382	281	211	431	249	451	375	316	1	279
158	000	54	261	156	89	302	109	311	239	172	1	173
163	54	000	233	130	59	279	98	298	222	165	1	160
382	261	233	000	114	183	68	165	125	74	137	1	164
281	156	130	114	000	80	160	66	183	105	81	1	132
211	89	59	183	80	000	231	59	251	173	123	1	141
431	302	279	68	160	231	000	205	80	78	156	1	184
249	109	98	165	66	59	205	000	213	140	76	1	135
451	311	298	125	183	251	80	213	000	88	149	1	201
375	239	222	74	105	173	78	140	88	000	88	1	152
316	172	165	137	81	123	156	76	149	88	000	1	145
1	1	1	1	1	1	1	1	1	1	1	0	1

$$(2.47)$$

The system of linear equations has been calculated by a computer program, and we obtain the following weights:

$$\Phi_1 = 0.090$$
$$\Phi_2 = 0.114 \qquad h'_n = \Sigma \Phi_i h_i = 24.6 \, \text{mm} \qquad\qquad (2.48)$$
$$\Phi_3 = 0.111$$
$$\Phi_4 = 0.088 \qquad \sigma^2 = \Sigma \Phi_i \Gamma(i,n) + \Phi - \Gamma(n,n)$$
$$\Phi_5 = 0.096$$
$$\Phi_6 = 0.060 \qquad\quad = 171.5 + 3.5 - 169.6$$
$$\Phi_7 = 0.118$$
$$\Phi_8 = 0.088 \qquad \sigma = 2.2 \qquad\qquad\qquad\qquad (2.49)$$
$$\Phi_9 = 0.095$$
$$\Phi_{10} = 0.086$$
$$\Phi_{11} = 0.053$$

$$\Theta = 3.59.$$

The calculation must be repeated for each considered time interval to obtain a time distribution of areal precipitation.

Advantages of the method:
The kriging technique provides estimated mean areal precipitation values
with an assessment of the estimation error.
Disadvantages of the method:
The selection of a theoretical semi-variogram model seems to be a more
arbitrary operation than a conclusive inference from the available records.
Aid to calculation:
The calculation can be made more easily by using the computer program
PMSB5 (Appendix 5).

2.2   Baseflow and Runoff

The effect of a rainfall event in a catchment area is registered by runoff
gauging stations. The station is normally equipped with a water-level
recorder which indicates stage versus time. Such a chart is called a stage
hydrograph. The stage hydrograph is transformed into a discharge
hydrograph by use of a rating curve.
The discharge hydrograph represents a flood wave and consists of different
components such as direct surface runoff, interflow, groundwater or base
flow, and river precipitation. The interflow, a subsurface flow reaching the
river in a relatively short time, and the river precipitation are commonly
considered parts of the direct surface runoff. Consequently, the discharge
hydrograph consists mainly of two components, the groundwater flow and
that of the surface.
The base flow consists of water that percolates downward until it reaches
the groundwater flow and flows to the river as groundwater discharge.
Groundwater accretion resulting from a particular rainfall event is normally
released over an extended period of time. Thus, groundwater sustains the
river flow during a period between precipitation events, and the base flow
hydrograph shows only a small increase during the storm.
The mathematical modelling of the rainfall-runoff process makes use of
time functions of precipitation and runoff tied together by an operator.
These functions have to correspond not only in a qualitative but also in
a quantitative manner. This means that the values of the integral of the functions
over a certain time interval have to be equal. In other words, the rainfall amount
of an event has to be equal to the discharge amount if there is no storage in
the watershed.
As the base flow is a long-time component in the discharge hydrograph,
the surface flow corresponds quantitatively with a part of the areal
precipitation; this latter is called effective rainfall, and the corresponding
surface flow is direct runoff.

Methods for Base Flow Separation

The simplest technique of base flow separation is to draw a straight line in the discharge hydrograph from the point at which the surface runoff begins to an intersection with the hydrograph recession. The starting point of surface runoff can be determined relatively clearly because of an evident increase in the curve, but the determination of the inter-section in the recession is a subjective procedure. The difficulty arises as a consequence of the growing contribution of groundwater flow in the discharge hydrograph.

One approach to avoid the influence of subjectivity is to plot the discharge hydrograph in a logarithmic scale system and to look for a break point at the end of the discharge curve.

Because of the exponential decrease of the discharge curve after the end of the rainfall event, the hydrograph shows a straight line in the logarithmic scale. The exponential decreasing factor for surface flow and groundwater flow is different. The consequence of this is that in a logarithmic scale the straight lines have different slopes, and the break point indicates the end of the surface runoff.

If this point is determined, a straight line can be drawn to the starting point of the event separating the base flow (Q1) from surface flow (Q2).

However, the straight line is also an approximation, because the accurate form of the separation line remains unknown.

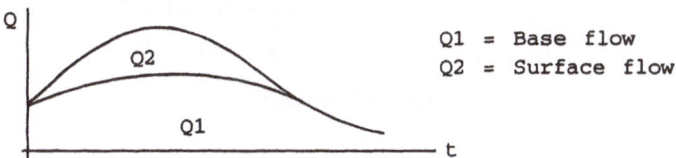

Q1 = Base flow
Q2 = Surface flow

Fig. 2.8 Baseflow separation

The amount of surface runoff can be calculated by the difference between integrated curves of base flow and surface flow, i.e. the determination of the area of surface flow.

In practice, a planimeter can be used, or if the chart is time-digitized, a computer program can be used for summation of differences.

2.3 Effective Rainfall

The rain gauge network provides a measure of time distribution of total rainfall in a watershed. Only one part of that precipitation, called net precipitation or effective rainfall, contributes directly to the surface runoff.

The other part of the total measured rainfall is composed of losses to interception, evaporation, depression storage, and infiltration.

The sum of losses of total measured rainfall is in the range of 25% − 80% for temperate zones of the earth, and in tropical zones values about 90% or even more can be obtained. Moreover, it is possible that after a rainfall event no surface discharge occurs.

All commonly used methods to determine the time distribution of effective rain are approximations, as many factors of influence cannot be caught by measurements at all.

Φ-Index Method

By this method the sum of rainfall losses and its time of development are constant during the storm period.

This is a rough approximation, because the time development of losses in nature is not constant.

The Φ-index represents the average rainfall intensity above which the volume of rainfall is equal to the volume of surface runoff.

Without storage from the equation of continuity it follows that:

$$\Sigma(P_i - \Phi_i) = \Sigma P_{eff} = \Sigma Q_{dir} \tag{2.50}$$

P_i = Mean areal precipitation in the time interval i

Φ_i = Constant rainfall loss in the time interval i

P_{eff} = Effective rain in the time interval i

Q_{dir} = Surface runoff in the time interval i.

Example for demonstrating the method:

Given problem:

Volume of mean areal precipitation: $\Sigma P_i = 75\,mm$

Time distribution of areal precipitation of an event:

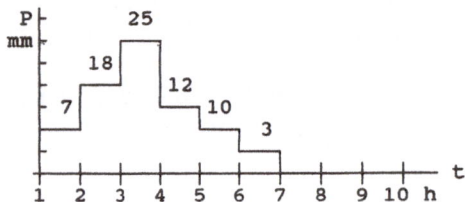

Fig. 2.9 Time distribution of mean areal precipitation

Volume of surface runoff: $\Sigma Q_i = 330\,m^3$

Catchment area: $A = 10,000\,m^2$

Desired: Effective rain distribution

Solution:
Conversion of $Q(m^3)$ in $Q(mm)$:

$$1\,m^3 = 1000\,l; \quad 1\,l/m^2 = 1\,mm \tag{2.51}$$

$$Q = 330\,m^3 = 330{,}000\,l \tag{2.52}$$

$$Q = 330{,}00\,l/10{,}000\,m^2 = 33\,l/m^2 = 33\,mm. \tag{2.53}$$

Estimation of a loss quota of Φ and verification of the continuity condition:

1. trial: $\Phi_i = 5\,mm/h$ estimated
   Verification: $7 - 5 =\ 2$
   $18 - 5 = 13$
   $25 - 5 = 20$
   $12 - 5 =\ 7$
   $10 - 5 =\ 5$
   $3 - 5 =\ -$

   $\Sigma\ 47\,mm > 33\,mm$

2. trial: $\Phi_i = 8\,mm/h$ estimated
   Verification: $7 - 8 =\ -$
   $18 - 8 = 10$
   $25 - 8 = 17$
   $12 - 8 =\ 4$
   $10 - 8 =\ 2$
   $3 - 8 =\ -$

   $\Sigma\ 33\,mm = 33\,mm$

Result: For the loss quota is obtained $\Phi_i = 8\,mm/h$
Illustrative representation of the result:

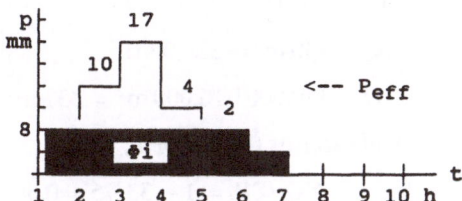

Fig. 2.10 Time distribution of the effective rain

Aid to calculation:
The calculation can be made more easily by using the computer program PREF1
(Appendix 6).

Method of Percentage Loss Rates

This method is similar to the Φ-index method, but the loss rates for each time interval depend on the rainfall depths by a constant factor, k.

$$\Phi_i = k \cdot P_i, \quad \text{with } k \in R.$$

It follows that:

$$\Sigma(P_i - \Phi_i) = \Sigma Q_i$$
$$\Sigma(P_i - kP_i) = \Sigma Q_i$$
$$\Sigma P_i(1 - k) = \Sigma Q_i$$
$$(1 - k) = \Sigma Q_i / \Sigma P_i$$
$$k = 1 - \Sigma Q_i / \Sigma P_i.$$

Example to demonstrate the method:

Given problem:

Volume of areal precipitation: $\Sigma P_i = 75$ mm

Time distribution of the mean areal precipitation of an event:

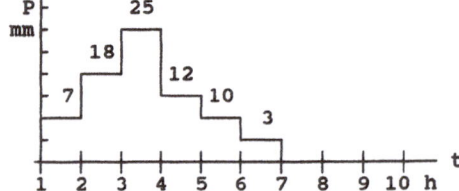

Fig. 2.11 Time distribution of the mean areal precipitation

Volume of surface runoff: $\Sigma Q_i = 330$ m^3

Catchment area: $A = 10,000$ m^2

Desired: Effective rain distribution

Solution:

Conversion of Q(m^3) in Q(mm):

$$1\,m^3 = 1000\,l; \quad 1\,l/m^2 = 1\,mm$$

$$Q = 330\,m^3 = 330,000\,l$$

$$Q = 330,000\,l/10,000\,m^2 = 33\,l/m^2 = 33\,mm$$

Calculation of the factor k:

$$k = 1 - \Sigma Q_i / \Sigma P_i = 1 - 33/75 = 0.56$$

Verification: $\begin{aligned} 7 - \;\;3.92 &= \;\;3.08 = a \\ 18 - 10.08 &= \;\;7.92 = b \\ 25 - 14.00 &= 11.00 = c \\ 12 - \;\;6.72 &= \;\;5.28 = d \\ 10 - \;\;5.60 &= \;\;4.40 = e \\ 3 - \;\;1.68 &= \;\;1.32 = f \end{aligned}$

$$\overline{}$$

$$\Sigma\;33.00\,mm = 33\,mm.$$

Illustrative representation of the result:

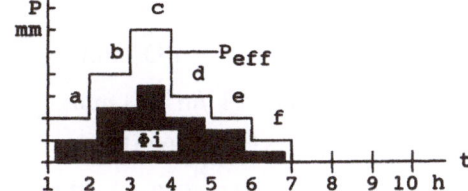

Fig. 2.12 Time distribution of the effective rain

Aid to calculation:
The calculation can be made more easily by using the computer program PREF2 (Appendix 7)

Method of Exponential Decreasing Loss Rates (Horton)
If infiltration is a dominating factor of the loss rate an exponential decreasing loss rate function (Horton function) can be used for determining an effective rain distribution.
Accordingly, it follows:

$$c(t) = kc + (ko - kc)\exp(-\alpha t) \tag{2.54}$$

$c(t)$ = Loss rate capacity (depths/time interval t)
 α = A constant factor specifying the rate of decrease
 kc = Permanent loss rate capacity
 ko = Initial loss rate capacity

From the condition of continuity it follows that:

$$\Sigma[P_i - c(t_i)] = \Sigma Q_i. \tag{2.55}$$

The parameters α, ko, kc can be determined by soil examination or estimated by means of soil maps.
Example demonstrating the method:
Given problem:
  Volume of the mean areal precipitation: $\Sigma P_i = 75$ mm
  Time distribution of the mean areal precipitation of an event

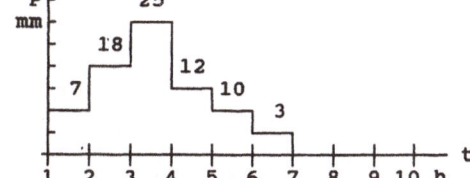

Fig. 2.13 Time distribution of the mean areal precipitation

Volume of surface runoff: $\Sigma Q_i = 330\,m^3$
Catchment area: $A = 10,000\,m^2$
Desired: Effective rain distribution
Solution:
Conversion of $Q(m^3)$ in $Q(mm)$:

$$1\,cbm = 1000\,l; \quad 1\,l/m^2 = 1\,mm$$

$$Q = 330\,m^3 = 330,000\,l$$

$$Q = 330,000\,l/10,000\,m^2 = 33\,l/m^2 = 33\,mm.$$

Parameter values of α and k are obtained from a soil map:

$$\alpha = 0.2, ko = 13.0, kc = 11.0$$

Verification: $7 - [2 + 11 \exp(-0.2 . 1)] = \div \quad = a$
$\qquad\qquad 18 - [2 + 11 \exp(-0.2 . 2)] = \ 9.00 = b$
$\qquad\qquad 25 - [2 + 11 \exp(-0.2 . 3)] = 16.85 = c$
$\qquad\qquad 12 - [2 + 11 \exp(-0.2 . 4)] = \ 4.63 = d$
$\qquad\qquad 10 - [2 + 11 \exp(-0.2 . 5)] = \ 3.33 = e$
$\qquad\qquad\ 3 - [2 + 11 \exp(-0.2 . 6)] = \div \quad = f$

$$\Sigma\ 34.5\,mm \approx 33\,mm.$$

Illustrative representation of the result:

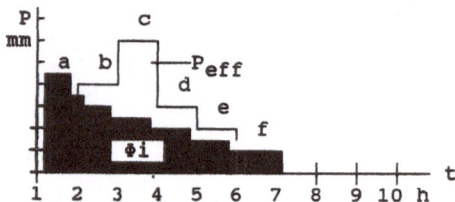

Fig. 2.14 Time distribution of the effective rain

Aid to calculation:
The calculation can be made more easily by using the computer program PMSB3 (Appendix 8).

3 Treatment of Hydrological Variables in Linear System Models

3.1 System-Theoretical Elements

The system-theoretical approach in hydrology is mainly applied to model the relationship between effective rain and surface runoff. It follows as a consequence that too many variables are involved in this process of transformation.

The question is aimed only at the effects caused by an action on a system. The inner structure, as well as the organization of the system, are not concerned.

The theory is thus sometimes called the "black box" theory. It is of interest to describe physical characteristics in terms of the mathematical relations governing a system.

Before dealing with properties of systems and their mathematical considerations, it is necessary to explain some terms used in the text.

The System
As already mentioned in Chapter 1, a system is like a field where the transition from input to output information is carried out. The system is a transformator consisting of a collection of interconnected components. It is characterized by an operator (H) indicating in which manner the transformation will be realized.

The inputs and outputs of the system:
Inputs and outputs represent physical quantities varying with time, and are therefore represented by time functions. These functions are also called signals. The output function is often called the response of the system.

The operator:
The operator is a generalization of the function term, and assigns a set of y-functions to a set of x-functions. It stands for a transformation function for inputs in outputs, and its form is significant for the behaviour of the system.

The transfer of system-theoretical terms to hydrological problems leads to the following identities:

System-input information = Effective rainfall $P(t)$
System   = Catchment area/soil $H[P(t)]$
System-output information = Surface runoff $Q(t)$

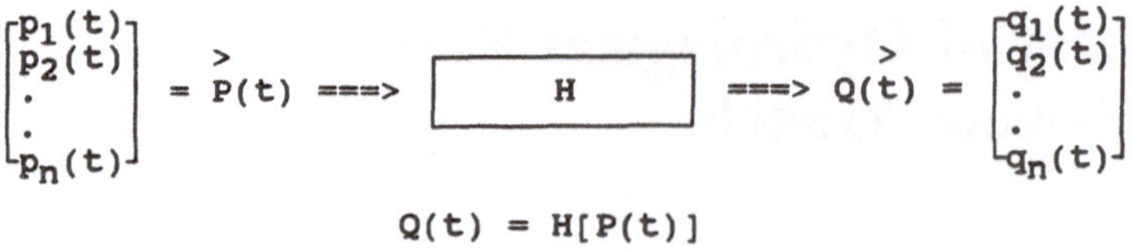

$$\begin{bmatrix} p_1(t) \\ p_2(t) \\ \cdot \\ \cdot \\ p_n(t) \end{bmatrix} = P(t) \overset{>}{===>} \boxed{\quad H \quad} \overset{>}{===>} Q(t) = \begin{bmatrix} q_1(t) \\ q_2(t) \\ \cdot \\ \cdot \\ q_n(t) \end{bmatrix}$$

$$Q(t) = H[P(t)]$$

Fig. 3.1 Schematic representation of a hydrological system

Examples of systems:

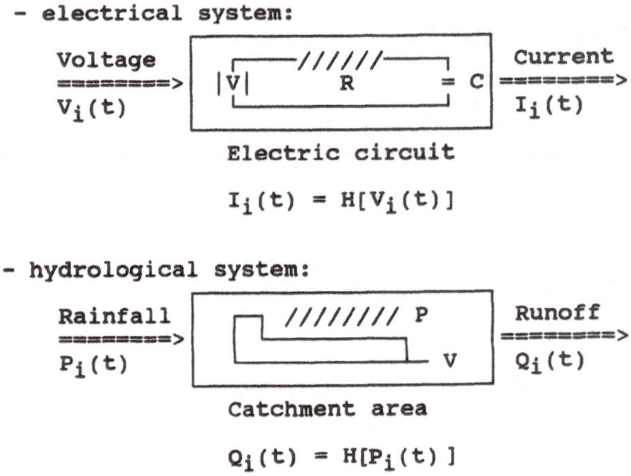

- electrical system:

$$I_i(t) = H[V_i(t)]$$

- hydrological system:

$$Q_i(t) = H[P_i(t)]$$

Fig. 3.2 Schematic representation of examples of systems

Depending on their properties, systems can be classified in several ways:

– stationary and non-stationary systems
– causal and non-causal systems
– linear and non-linear systems
– deterministic and stochastic systems
– stable and unstable systems
– continuous-time and discrete-time systems

Some classes of systems are of particular interest in hydrology, and it is therefore useful to consider them in more detail.

Continuous-time and Discrete-Time Systems
If the time functions of inputs and outputs are considered to be functions of a continuous variable, i.e. if the system is able to give a reaction in any instant of time, it is called a continuous-time system. The functions are written as $p(t)$ or $q(t)$.

The input and output functions are those of a continuous variable, but they do not have to be necessarily continuous functions in a mathematical sense.
If the system reacts only at discrete instants and values of the functions, or if the system between these instants is unknown, constant, or of no interest, the system is called a discrete-time one. The functions are written as p(n) or q(n), where n stands for a discrete time interval.

Stationary Systems

A system is called a stationary system if its properties of input-output relations are time invariant, i.e. they do not change with time. This means that the results of system analysis are time independent, and can therefore be used directly for model application at any time.
In the case of hydrological systems and their stationarity, normally no time independence of the behaviour is given owing to the influence of seasons, climates, periods, etc. In this text dealing with an approximation of system behaviour for hydrological systems, time independence is assumed.

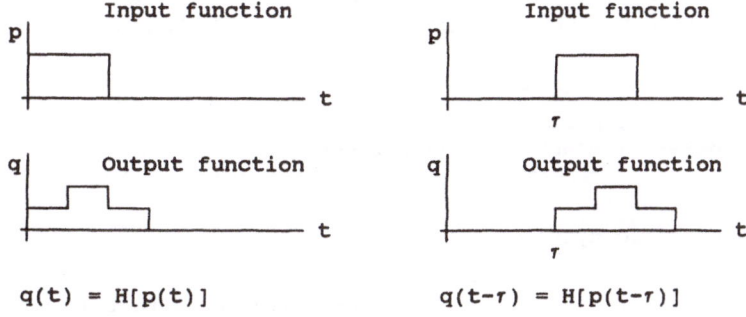

Fig. 3.3 Time stationarity of inputs and outputs

The condition for stationarity can be formulated mathematically is follows:
The system having the input-output relation $q(t) = H[p(t)]$ is time invariant or stationary if and only if

$$q(t - \tau) = H[p(t - \tau)]$$
(3.1)

for any q(t) and any τ.

Causal Systems

A system is called causal if its output function, caused by that of input function, does not depend on future values of input.
Causal systems are physical systems, and for that reason constraints on the mathematical formulation of the problem must be taken into consideration.

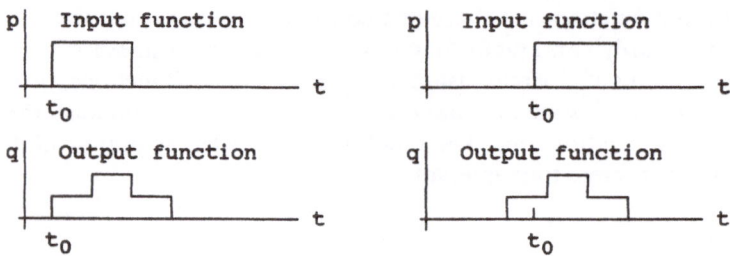

Fig. 3.4 Causal and non-causal system outputs

Mathematical formulation of conditions for:

Causal system	noncausal system
$q(t < t_0) = p(t < t_0) \quad = 0$	$q(t < t_0) \neq p(t < t_0) \quad \neq 0$
$\quad = H[p(t < t_0)] = 0$	$\neq H[p(t < t_0)] \neq 0.$

The black box model serves as a cybernetical description of causal relations. It is applied to both deterministic and stochastic modelling. In stochastic modelling, different output functions are caused by the same input function. The probability of the output is caused only by the input.

```
Deterministic system:

Cause--->| Black box |--->Effect
```

```
Stochastic system   :
```

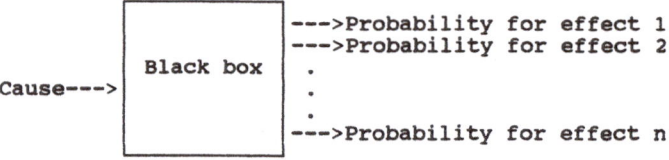

Fig. 3.5 Deterministic and stochastic systems

Linear Systems

A system is called linear if it satisfies the principle of superposition; this means, that the application of an operator to a sum of input information providing an output information is equivalent to the sum of single output information separately produced by the operator. Mathematically formulated it follows that: a system is linear if and only if

$$H[\alpha P_1(t) + \beta P_2(t)] = \alpha H[P_1(t)] + \beta H[P_2(t)], \tag{3.2}$$

where α, β are constants.

In the case where $\alpha = \beta = 1$ we obtain:

$$H[P_1(t) + P_2(t)] = H[P_1(t)] + H(P_2(t)].\tag{3.3}$$

This means that linear systems are additive.
In the case where $P_1(t) = 0$ or $P_2(t) = 0$ we obtain:

$$H[\alpha P(t)] = \alpha H(P(t)].\tag{3.4}$$

This means that linear systems are homogeneous.
Conversely, a system is linear if and only if it is both additive and homogeneous.
The principle of superposition can be specified to:

- Superposition of amplitudes:

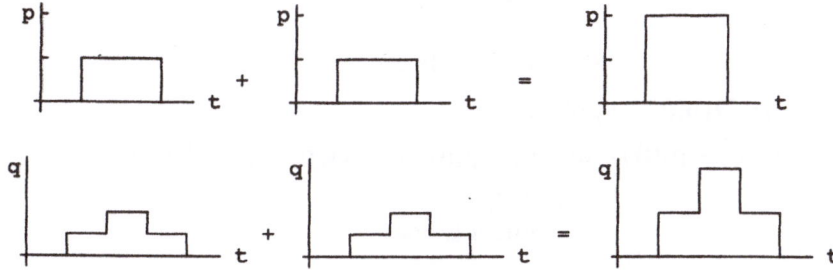

Fig. 3.6 Superposition of amplitudes

- Superposition of time:

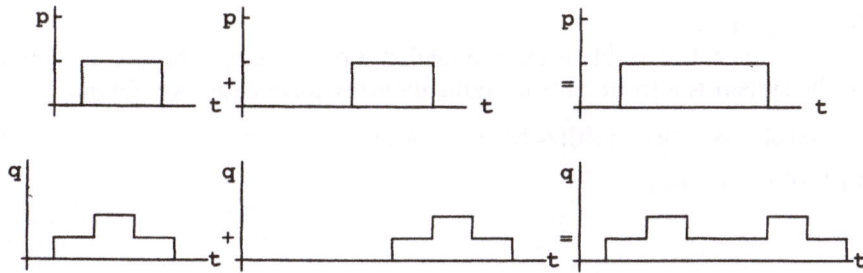

Fig. 3.7 Superposition of time

Some examples are given to make the reader well versed in the meaning of linear and non-linear system characteristics.

1. $q(t) = H[p(t)] = p^2(t)$ \hfill (3.5)

$$H[p_1(t) + p_2(t)] = [p_1(t) + p_2(t)]^2 \neq H[p_1(t)] + H[p_2(t)] = p_1^2(t) + p_2^2(t)$$
$$\Rightarrow \text{the system is non-linear.}$$

2. $q(t) = H[p(t)] = \alpha p(t) + \beta : \alpha, \beta \in R$ (3.6)

$$H[p_1(t) + p_2(t)] = \alpha[p_1(t) + p_2(t)] + \beta \neq Hp_1(t) + Hp_2(t)$$
$$= \alpha p_1(t) + \alpha p_2(t) + 2\beta$$
$$\Rightarrow \text{the system is non-linear, if } \beta \neq 0.$$

In the case that $\beta = 0$, it follows: the system is linear.

3. $q(t) = H[p(t)] = \int_0^t p(\tau)\,d\tau$ (3.7)

$$H[p_1(t) + p_2(t)] = \int_0^t [p_1(\tau) + p_2(\tau)]\,d\tau = H[p_1(t)] + H[p_2(t)]$$

$$= \int_0^t p_1(\tau)\,d\tau + \int_0^t p_2(\tau)\,d\tau$$

$$\Rightarrow \text{the system is linear.}$$

4. $q(t) = H[p(t)] = dp(t)/dt$ (3.8)

$$H[p_1(t) + p_2(t)] = d[p_1(t) + p_2(t)]/dt = H[p_1(t)] + H[(p_2(t)]$$
$$= p_1'(t) + p_2'(t)$$
$$\Rightarrow \text{the system is linear.}$$

Deterministic Systems

A system is called deterministic if the functions involved in the system are deterministic; this means that no random functions or random variables are included in the consideration.

Stable Systems

A system is called stable if for a bounded input function the output function of the system is also bounded. Mathematically formulated we obtain:

$$|p(t)| < N \quad \text{and} \quad |q(t)| < M \qquad (3.9)$$

with N, M as constants.

3.2   Concepts in Linear Systems

The behaviour of a system exposed to external influences or loads is characterized by its operator determining the reaction. Only with a known operator is it possible to predetermine the reaction of a system with given or predicted loadings. Therefore, in the analysis of linear systems we try to obtain sufficient information for the determination of the system-operator.

The behaviour of a system is usually described by mathematical formulas composed of differential equations. The system-operator consists of a set of such equations (lumped linear systems). The form of equations might be the following:

$$ap(t) + bp'(t) + cp''(t) + \cdots = \alpha q(t) + \beta q'(t) + \cdots \qquad (3.10)$$

(ordinary differential equation with constant coefficients).

To propose any such equation, it is normally necessary to have information about the inner structure of the system describable by differentials.

If the type of differential equation is known, the parameters must be determined by adjustment methods.

In linear systems, two basic approaches are known to determine the system operator:

– direct approach and
– indirect approach

The direct approach consists in solving the differential equation – the type of which is assumed to be known – for time functions and a known set of initial conditions. By using one of the standard methods, the problem of parameter adaptation in a particular solution of the differential equation will arise. The resulting time functions are usually analyticals.

The problem of which type of differential equation is assumed to conform to the behaviour of the system, often causes considerable difficulties.

For this reason, in hydrology simple models are frequently used to obtain the physical performance of the system behaviour. In the case of a linear reservoir model the following differential equation governing the system is obtained:

$$dq(t)/dt + \alpha q(t) = \alpha p(t). \qquad (3.11)$$

This equation can be solved relatively easily, and depending on the initial conditions, the parameter α can be determined. However, the adaptation of the model to a real physical system is difficult because the model represents only specific characteristics of a linear reservoir and is not representative of the system behaviour of a watershed.

$$\begin{bmatrix} p_1(t) \\ p_2(t) \\ \cdot \\ \cdot \\ p_n(t) \end{bmatrix} = \overset{>}{P}(t) \implies \boxed{Q = H(P, P', P'', \ldots)} \implies \overset{>}{Q}(t) = \begin{bmatrix} q_1(t) \\ q_2(t) \\ \cdot \\ \cdot \\ q_n(t) \end{bmatrix}$$

Fig. 3.8 Lumped linear system and direct approach to linear system analysis

The indirect approach consists in searching for a differential equation for time functions resolved in elementary components. These components are the simple functions u(t), v(t) arising, for example, from an expansion of a periodic function into a Fourier series of sines and cosines or exponentials. Step functions or impulse functions can equally well be used. The response of the system to such elementary functions reflects the system behaviour. Based on its linearity, the output function is obtained by addition of the elementary response functions. For linear systems it is easier to obtain information on the system behaviour by using elementary functions than to use complete time functions.

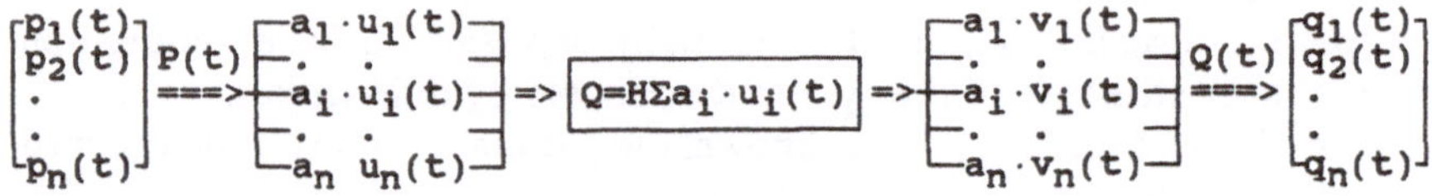

$$
\begin{bmatrix} P_1(t) \\ P_2(t) \\ \cdot \\ P_n(t) \end{bmatrix} \overset{P(t)}{===>} \begin{bmatrix} a_1 \cdot u_1(t) \\ \cdot \quad \cdot \\ a_i \cdot u_i(t) \\ \cdot \quad \cdot \\ a_n \ u_n(t) \end{bmatrix} => \boxed{Q=H\Sigma a_i \cdot u_i(t)} => \begin{bmatrix} a_1 \cdot v_1(t) \\ \cdot \quad \cdot \\ a_i \cdot v_i(t) \\ \cdot \quad \cdot \\ a_n \cdot v_n(t) \end{bmatrix} \overset{Q(t)}{===>} \begin{bmatrix} q_1(t) \\ q_2(t) \\ \cdot \\ q_n(t) \end{bmatrix}
$$

Fig. 3.9 Indirect approach to linear system analysis

$$P(t) = \Sigma a_i \cdot u_i(t) \tag{3.12}$$
$$Q(t) = H[P(t)] = H[\Sigma a_i \cdot u_i(t)] = \Sigma a_i \cdot H(u_i(t)) = \Sigma a_i \cdot v_i(t).$$

Responses of the system to elementary functions (system operator) is given by the following relationship:

$$v_i(t) = H[u_i(t)]$$

a_i = the amplitude of elementary functions.

In the following the indirect approach is considered in detail. The resulting mathematical methods are used to calculate the system behaviour of the rainfall-runoff relationship.

In the beginning there is the question of which type of elementary functions is suitable to decompose time functions of precipitation and runoff.

The decomposition of analytical functions is preferably done with functions mathematically simple in handling. For this reason, potential elementary functions are, for example, impulse functions (delta function) $\delta(t)$, step functions $\mu(t)$, or exponential functions $\exp(s)$.

The decomposition in delta functions or step functions can be developed relatively easily and in this procedure the independent time variable (t) is maintained. The field of application for this type of function is called time domain. The decomposition in exponential functions can be carried out by the Laplace transformation method, and in this procedure the independent time variable (t) appears in the form of a complex number representing any frequency. The field for such applications is called the frequency domain.

A summary of the methods mentioned above is given in the following table.

Table 3.1. Summary of the methods. (Schwarz and Friedland 1965)

Procedure	Time domain		Frequency domain
Resolution of input	Resolution into unit impulses	Resolution into unit steps	Resolution into complex exponentials
	$\delta(t)$	$\mu(t)$	$\exp(s)$
Response to component	Impulse response	Step response	System transfer function
	$h(t)$	$a(t)$	$H(s)$
Composition	Superposition/convolution		Multiplication/inversion

3.3 Treatment of Variables in the Time Domain

3.3.1 Impulse and Step Function

Impulse or Delta Function
a) Continuous time variables
The meaning of the delta function is given by the following illustrative figure:

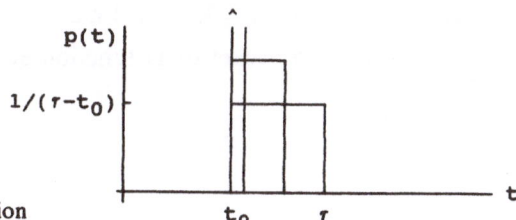

Fig. 3.10 Representation of delta function

Proceeding from a pulse function $p(t)$ with:

$$p(t) = 0, \qquad \text{if} \quad \tau < t < t_0 \tag{3.13}$$
$$p(t) = 1/(\tau - t_0), \quad \text{if} \quad t_0 < t < \tau \tag{3.14}$$

the definition of the delta function is derived as follows:

$$\delta(t - t_0) = \lim_{\tau \to t_0} p(t) = 0, \quad \text{if} \quad t \neq t_0 \tag{3.15}$$

$$\delta(t - t_0) = \lim_{\tau \to t_0} p(t_0) \neq 0, \quad \text{if} \quad t = t_0. \tag{3.16}$$

The function $\delta(t - t_0)$ is called the Dirac delta function or Dirac impulse.
The function is zero for all values of t with $t \neq 0$, and for $t = t_0$ the function rises to infinite.
Considering the vicinity of t_0, for values of τ arbitrarily narrow to t_0 the accordingly obtained values of the function are arbitrarily high.

The integration of the delta function provides:

$$\int_{-\infty}^{+\infty} \delta(t-t_0)\,dt = 1 \qquad \int_{-\infty}^{+\infty} \delta(t-t_0)\,dt_0 = 1 \tag{3.17}$$

Proof:

$$t \neq t_0 : \Rightarrow \delta(t-t_0) = 0 \Rightarrow \int_{-\infty}^{t_0} \delta(t-t_0)\,dt + \int_{\tau}^{+\infty} \delta(t-t_0)\,dt = 0 \tag{3.18}$$

$$t = t_0 : \Rightarrow \delta(t-t_0) = \delta(0) = \lim_{\tau \to t_0} p(t_0) = \lim_{\tau \to t_0} 1/(\tau - t_0) \tag{3.19}$$

$$\int_{t_0}^{\tau} \delta(t-t_0)\,dt = \lim_{\tau \to t_0} \int_{t_0}^{\tau} 1/(\tau - t_0)\,dt = \lim_{\tau \to t_0} 1/(\tau - t_0) \int_{t_0}^{\tau} dt \tag{3.20}$$

$$= \lim_{\tau \to t_0} (\tau - t_0)/(\tau - t_0) = 1.$$

b) Discrete time variables

In the use of discretized time variables, the impulse function is explained similarly to that of the continuous ones. The difference is that for the position $n = k$, the function is one and does not take on infinite values.

Definition of the delta function: (Kronecker delta function)

$$\delta(n-k) = 0, \quad \text{if} \quad n \neq k \tag{3.21}$$

$$\delta(n-k) = 1, \quad \text{if} \quad n = k \qquad n, k \in Z. \tag{3.22}$$

The sum of the Kronecker delta function accordingly is:

$$\sum_{k=-\infty}^{k=+\infty} (n-k) = 1 \tag{3.23}$$

Proof:

$$k \neq n \Rightarrow \delta(k-n) = 0 \tag{3.24}$$

$$k = n \Rightarrow \delta(0) \quad = 1. \tag{3.25}$$

Step Function

a) Continuous time variables

The step function is defined by considering the following illustrative figure:

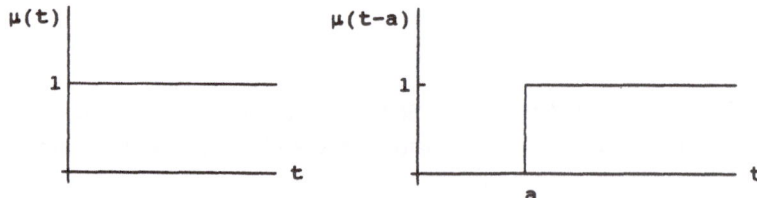

Fig. 3.11 Unit step

Definition of the step function:

$$\mu(t) = 0, \qquad \text{if } t \le 0 \tag{3.26}$$

$$\mu(t) = 1, \qquad \text{if } t > 0 \tag{3.27}$$

$$\mu(t - a) = 0, \quad \text{if } t \le a \tag{3.28}$$

$$\mu(t - a) = 1, \quad \text{if } t > a. \tag{3.29}$$

The derivative of the step function is given by the delta function.

$$d\mu(t)/dt = \delta(t) \tag{3.30}$$

or:

$$\int_{-\infty}^{t} \delta(\alpha - \tau)d\alpha = \mu(t - \tau). \tag{3.31}$$

Illustration:
The step function may change from zero to one in the finite time interval τ with a constant slope and then the derivative is formed. It is shown in the following figure that by diminution of the time interval τ to zero the impulse function results as derivative of the step function.

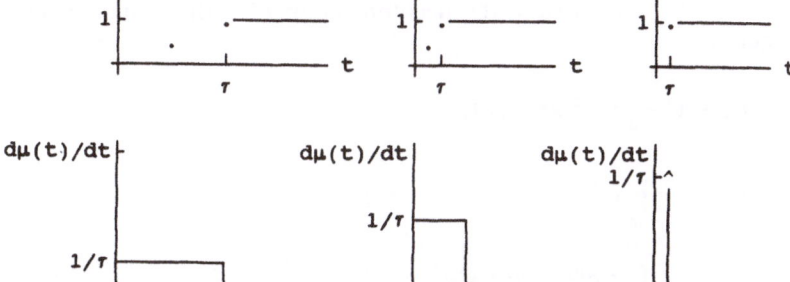

Fig. 3.12 Derivative of the unit step function

b) Discrete time variables
By analogy with continuous time variables, the step function can be formed as follows:
Definition of the step function:

$$\mu(n) = 0, \qquad \text{if } n \le 0 \tag{3.32}$$

$$\mu(n) = 1, \qquad \text{if } n > 0 \tag{3.33}$$

$$\mu(n - k) = 0, \quad \text{if } n \le k \tag{3.34}$$

$$\mu(n - k) = 1, \quad \text{if } n > k \qquad \text{with } n, k \in Z. \tag{3.35}$$

3.3.2   Resolution of System Input Functions

Each time function $f(t)$ considered as a real system input function can be written
as a sum of elementary functions, like step functions $\mu(t)$ or delta functions $\delta(t)$.
a) Continuous time variables
   Decomposition into impulse functions $\delta(t)$:

$$f(t) = f(t) \cdot \int_{-\infty}^{+\infty} \delta(t - t_0) dt_0 = \int_{-\infty}^{+\infty} f(t_0) \delta(t - t_0) dt_0 \tag{3.36}$$

with

$$\int_{-\infty}^{+\infty} \delta(t - t_0) dt_0 = 1 \tag{3.37}$$

or

$$f(t) = f(t) \cdot \int_{-\infty}^{+\infty} \delta(t - \tau) d\tau = \int_{-\infty}^{+\infty} f(\tau) \delta(t - \tau) d\tau \tag{3.38}$$

with

$$\int_{-\infty}^{+\infty} \delta(t - \tau) d\tau = 1. \tag{3.39}$$

Example to demonstrate the decomposition of a time function into impulse
functions:

$$f(t) = t^2 = \int_{-\infty}^{+\infty} \tau^2 \delta(t - \tau) d\tau \tag{3.40}$$

$$t = 1: \quad \int_{-\infty}^{+\infty} \tau^2 \delta(1 - \tau) d\tau = 1^2 \cdot 1 = 1 \tag{3.41}$$

$$t = 2: \quad \int_{-\infty}^{+\infty} \tau^2 \delta(2 - \tau) d\tau = 2^2 \cdot 1 = 4 \tag{3.42}$$

...

Decomposition into step functions $\mu(t)$:
By substitution for:

$$d\mu(t)/dt = \dot{\mu}(t) = \delta(t), \text{ resp.} \tag{3.43}$$

$$d\mu(t - \tau)/dt \cdot (1) = \dot{\mu}(t - \tau) = \delta(t - \tau) \tag{3.44}$$

$$d\mu(t - \tau)/d\tau \cdot (-1) = -\dot{\mu}(t - \tau) = -\delta(t - \tau) \tag{3.45}$$

we obtain:

$$f(t) = f(t) \cdot \int_{-\infty}^{+\infty} \dot{\mu}(t - t_0) dt_0 = \int_{-\infty}^{+\infty} f(t_0) \dot{\mu}(t - t_0) dt_0 \tag{3.46}$$

or

$$f(t) = f(t) \cdot \int_{-\infty}^{+\infty} \dot{\mu}(t - \tau) \, d\tau = \int_{-\infty}^{+\infty} f(\tau) \dot{\mu}(t - \tau) \, d\tau \tag{3.47}$$

From differentiation with respect to τ follows:

$$d[f(\tau)\mu(t - \tau)]/d\tau = -f(\tau)\dot{\mu}(t - \tau) + \dot{f}(\tau)\mu(t - \tau) \tag{3.48}$$

Finally, by integration we obtain:

$$f(t) = \int_{-\infty}^{+\infty} f(\tau)\dot{\mu}(t - \tau) \, d\tau = -|f(\tau)\mu(t - \tau) \mathop{|}_{-\infty}^{+\infty} + \int_{-\infty}^{+\infty} \dot{f}(\tau)\mu(t - \tau) \, d\tau \tag{3.49}$$

$$= f(-\infty)\mu(t + \infty) - f(+\infty)\mu(t - \infty) + \int_{-\infty}^{+\infty} \dot{f}(\tau)\mu(t - \tau) \, d\tau. \tag{3.50}$$

As the step of the function $\mu(t - a)$ with $a = \infty$ lies at infinity, it follows for $t < \infty$ that the step function is zero, $\mu(t - \infty) = 0$.
For time functions we are interested only in function values with $t > 0$, and therefore the function of $f(-t)$ or $f(-\infty)$ can be considered as zero.
Accordingly, the resolution in step functions will be:

$$f(t) = \int_{0}^{\infty} \dot{f}(\tau)\mu(t - \tau) \, d\tau. \tag{3.51}$$

Example for demonstrating the decomposition of a time function into step functions:

$$f(t) = t^2 = 2 \int_{0}^{\infty} \tau\mu(t - \tau) \, d\tau \tag{3.52}$$

$$t = 1: \quad 2 \int_{0}^{1} \tau\mu(1 - \tau) \, d\tau = 2 \cdot 1^2/2 \cdot 1 = 1 \tag{3.53}$$

$$t = 2: \quad 2 \int_{0}^{2} \tau\mu(2 - \tau) \, d\tau = 2 \cdot 2^2/2 \cdot 1 = 4 \tag{3.54}$$

...

b) Discrete time variables
Decomposition into impulse functions $\delta(n)$:
If we are dealing with discrete time functions given for precipitation in the form of histograms, the decomposition is made by the Kronecker delta function by analogy with continuous time functions as follows:

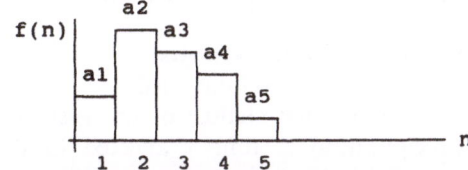

Fig. 3.13 Decomposition of a discrete time function into impulse functions

$$f(n) = \sum_{k=-\infty}^{k=+\infty} f(k)\delta(n-k) \tag{3.55}$$

$$\begin{aligned}
f(1) &= \Sigma f(k)\delta(1-k) = f(1)\delta(0) = f(1) = a1 \\
f(2) &= \Sigma f(k)\delta(2-k) = f(2)\delta(0) = f(2) = a2
\end{aligned}$$

$$\begin{aligned}
\cdot \quad\quad \cdot \quad\quad \cdot \quad\quad \cdot \cdot \cdot \\
\cdot \quad\quad \cdot \quad\quad \cdot \quad\quad \cdot \cdot \cdot
\end{aligned} \tag{3.56}$$

$$f(n) = \Sigma f(k)\delta(n-k) = f(n)\delta(0) = f(n) = an$$

Decomposition into step functions $\mu(n)$:

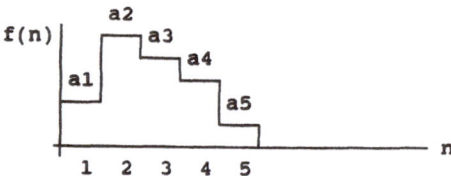

Fig. 3.14 Decomposition of a discrete time function into step functions

$$f(n) = \sum_{k=1}^{k=n} f(k)\mu(n-k) \tag{3.57}$$

$$\begin{aligned}
f(1) &= (a_1 - 0)/1\,\mu(1-1) & = a_1 \\
f(2) &= (a_1 - 0)/1\,\mu(2-1) + (a_2 - a_1)/1\,\mu(2-2) & = a_2 \\
f(3) &= (a_1 - 0)/1\,\mu(3-1) + (a_2 - a_1)/1\,\mu(3-2) + (a_3 - a_2)/1\,\mu(3-3) & = a_3
\end{aligned}$$

$$\begin{aligned}
\cdot \quad\quad \cdot \quad\quad\quad \cdot \quad\quad\quad \cdot \quad\quad\quad \cdot \\
\cdot \quad\quad \cdot \quad\quad\quad \cdot \quad\quad\quad \cdot \quad\quad\quad \cdot
\end{aligned}$$

$$f(n) \doteq \Sigma f(k)\mu(n-k) \qquad\qquad\qquad = a_n. \tag{3.58}$$

3.3.3 System Response Function

The theory of linear systems and their application in hydrology in principle requires dealing in two fields: analysis and synthesis.
The object of the analysis of the system is to determine the system operator, whereas synthesis tries to gain some perspectives by simulation of system behaviour with a known operator.
For system analysis with functions resolved in elementary functions, the system operator as a differential equation can be formulated in the following way:

Continuous Time Variables
Resolution in step functions:
An infinitesimal change of the system input function p(t) (time function of precipitation) at time τ affected to the system H causes a corresponding

infinitesimal change of the system output function q(t) (time function of runoff).

$$dq(t) = H[\mu(t - \tau) dp(\tau)]. \tag{3.59}$$

By integration of the differential equation we obtain:

$$q(t) = H\left\{ \int_0^{+\infty} \mu(t - \tau)[dp(\tau)/d\tau] \, d\tau \right\} = \int_0^{+\infty} \dot{p}(\tau)H[\mu(t - \tau)] \, d\tau. \tag{3.60}$$

For linear systems, the same result is given if step functions $\mu(t)$ of a resolved input function p(t) are exposed to the system behaviour H.
The system response $H[\mu(t - \tau)]$ to a step function is called step response function $a(t - \tau)$

$$a(t - \tau) = H[\mu(t - \tau)]. \tag{3.61}$$

For stationary and causal systems yields:

$$q(t) = \int_0^t \dot{p}(\tau)a(t - \tau) \, d\tau = \int_0^t \dot{p}(t - \tau)a(\tau) \, d\tau$$

Resolution in impulse functions:
If the impulse functions $\delta(t)$ of a resolved input function p(t) are exposed to the system behaviour H, the system output function q(t) is given as follows:

$$q(t) = H[p(t)] = H\left[\int_{-\infty}^{+\infty} p(\tau)\delta(t - \tau) \, d\tau \right] = \int_{-\infty}^{+\infty} p(\tau)H[\delta(t - \tau)] \, d\tau. \tag{3.62}$$

The system response $H[\delta(t - \tau)]$ to an impulse function is called the impulse response function $h(t - \tau)$

$$h(t - \tau) = H[\delta(t - \tau)]. \tag{3.63}$$

Then, for the output function holds:

$$q(t) = \int_{-\infty}^{+\infty} p(\tau)h(t, \tau) \, d\tau. \tag{3.64}$$

If this integral is considered in dependence on time t, it is called the convolution integral.
For $\tau > t$ and $t \geq 0$ the integral yields:

$$q(t) = \int_0^t p(\tau)h(t, \tau) \, d\tau. \tag{3.65}$$

The name convolution integral is used for calculating the function q(t), if the functions p(t) and h(t) are known. If the function h(t) is calculated and the functions q(t) and p(t) are known, the integral is called an integral equation of the Volterra type.

For stationary and causal systems this yields:

$$q(t) = \int_0^t p(\tau)h(t - \tau)d\tau = \int_0^t p(t - \tau)h(\tau)d\tau$$

Discrete Time Variables
Step function:
On the analogy of continuous time variables, the starting point for system response is the input function resolved in step functions.

$$p(n) = \sum_{-\infty}^{+\infty} \dot{p}(k)\mu(n - k).$$ (3.66)

For the system output function we obtain:

$$q(n) = H[p(n)] = \sum_{-\infty}^{+\infty} \dot{p}(k)H\mu(n - k) = \sum_{-\infty}^{+\infty} \dot{p}(k)a(n, k).$$ (3.67)

For stationary and causal systems this yields:

$$q(n) = \sum_0^n \dot{p}(k)a(n - k) = \sum_0^n \dot{p}(n - k)a(k)$$

Impulse function:
In analogy with continuous time variables it follows that:

$$p(n) = \sum_{-\infty}^{+\infty} p(k)\delta(n - k).$$ (3.68)

Then we obtain for the system output function:

$$q(n) = H[p(n)] = \sum_{-\infty}^{+\infty} p(k)H\delta(n - k) = \sum_{-\infty}^{+\infty} p(k)h(n, k).$$ (3.69)

For stationary and causal systems it results that:

$$q(n) = \sum_0^n p(k)h(n - k) = \sum_0^n p(n - k)h(k).$$

This formula can also be written as follows:

$$\begin{aligned}
q(0) &= h(0 - 0) \quad p(0) \\
q(1) &= h(1 - 0) \quad p(0) + h(1 - 1) \quad p(1) \\
q(2) &= h(2 - 0) \quad p(0) + h(2 - 1) \quad p(1) + h(2 - 2) \quad p(2)
\end{aligned}$$ (3.70)

Using the symbols of vectors and matrices

$$\begin{bmatrix} q(0) \\ q(1) \\ q(2) \\ \cdot \\ \cdot \end{bmatrix} = \overset{>}{q} \qquad \begin{bmatrix} h(0) & & \\ h(1) & h(0) & \\ h(2) & h(1) & h(0) \\ \cdot & & \cdot \\ \cdot & & \cdot \end{bmatrix} = \overline{h} \qquad \begin{bmatrix} p(0) \\ p(1) \\ p(2) \\ \cdot \\ \cdot \end{bmatrix} = \overset{>}{p}$$

the formula becomes:

$$\overset{>}{q} = \overline{h} * \overset{>}{p}.$$ (3.71)

Constraints for System Response

1. The continuity condition requires: $\int_0^\infty h(t)\,dt = 1$

2. For real discharges is required: $h(t) \geq 0$
   (negative values for discharge must be excluded).

In hydrology a system response function satisfying such conditions is called a unit hydrograph.

Relationship Between Step Response and Impulse Response
The derivative of the step function is equal to the impulse function, conversely considered the integrated impulse function is equal to the step function.

$$da(t)/dt = h(t) \quad \text{and} \quad a(t) = \int_0^t h(u)\,du.$$ (3.72)

Description of System Behaviour by Differential Equation
and Integral Equation
As mentioned above, the system behaviour can be described by a set of differential equations.
Considering the indirect approach, the input functions resolved in elementary functions produce any system response for output functions. The system behaviour is characterized by the system response. To obtain an explicit expression for the response function, the integral must be solved with known input and output functions.
If the system behaviour can be described by differential equations, the integral equation can be transformed into a differential equation, and the result of both equations is the same.
As is known, each differential equation can be transformed into an integral equation, but conversely each integral equation cannot be transformed into a differential equation. Consequently, integral equations can be considered a more comprehensive tool for description of linear systems, and those that can be described by differential equations form a special group among them.

3.3.4   Synthesis of System Output Functions by Convolution

The convolution process formulated by the convolution integral represents the synthesis of output functions. A graphical interpretation of it may be given in the following form:

a) continuous variables:

Example for graphical interpretation:

1. Given functions h(t),p(t):

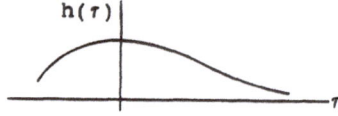   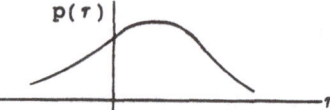

2. Representation of the "folding"-process:

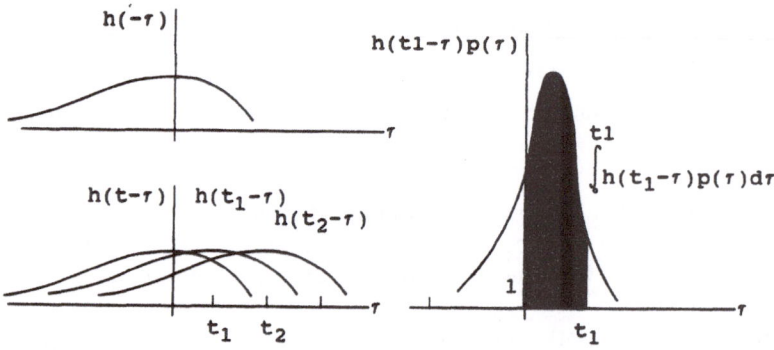

Fig. 3.15 Graphical representation of "folding"

The analytical calculation of the convolution integral is demonstrated by the following example:

Example to demonstrate the analytical calculation of the convolution:

Given: $h(t) = \sqrt{t} \exp(-t)$

$\qquad p(t) = t \exp(-t)$

Desired:

$$q(t) = \int_0^t h(\tau)p(t-\tau)\,d\tau$$

Solution:

$$q(t) = \int_0^t \sqrt{\tau}\exp(-\tau)\cdot(t-\tau)\exp - (t-\tau)\,d\tau = \exp(-t)\int_0^t \sqrt{\tau}\cdot(t-\tau)\,d\tau$$

$$q(t) = 4/15\cdot t^{5/2}\cdot \exp(-t)$$

b) Discrete variables:

For discrete variables the convolution process is graphically demonstrated by the following example:

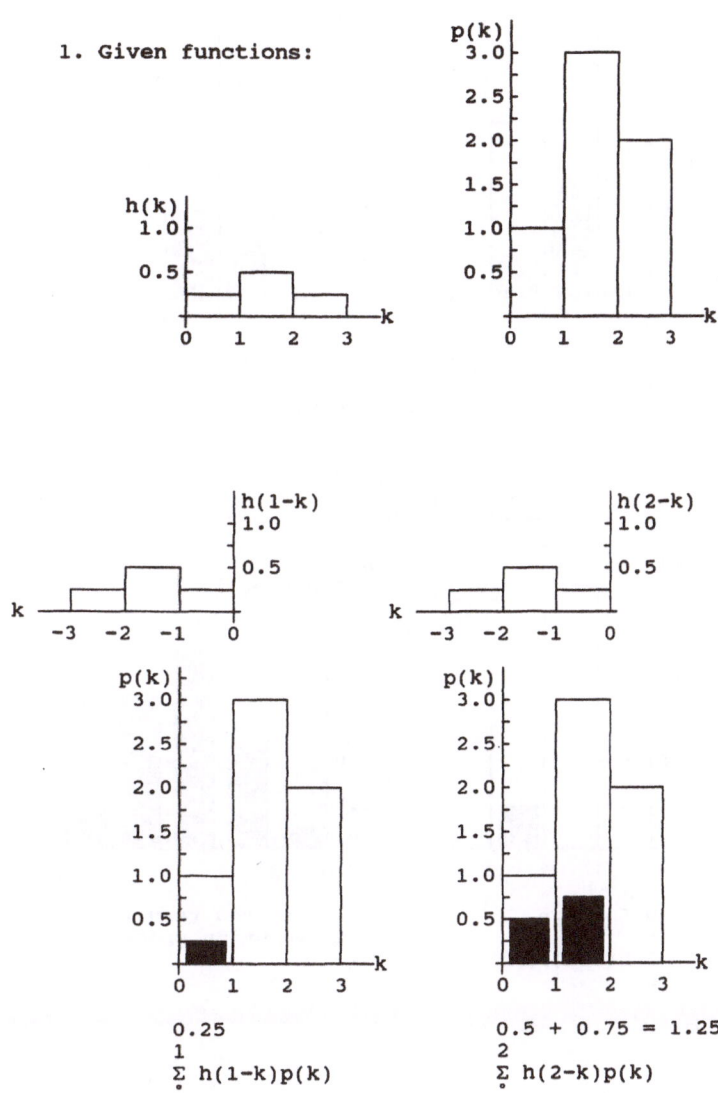

Fig. 3.16 Graphical representation of the convolution

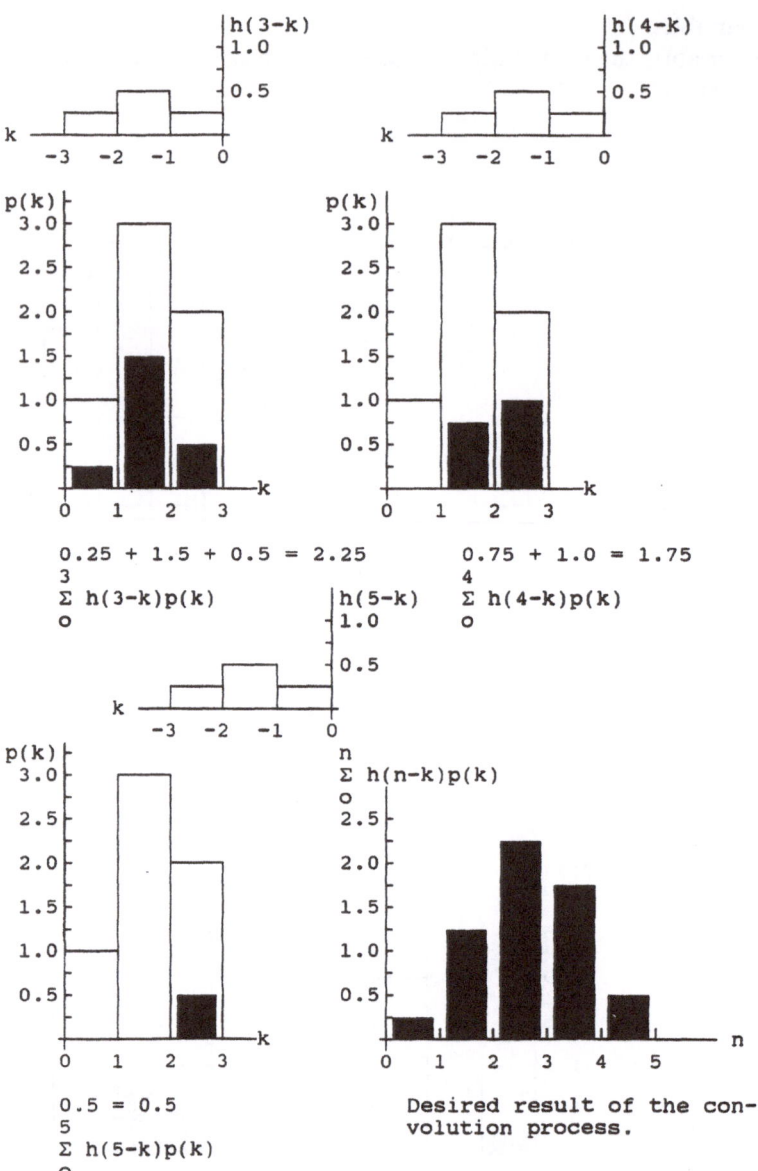

Fig. 3.16 (Continued)

Equally the convolution process can be considered as a "moving average" process.

The sum $q(n) = \sum_0^n p(k)h(n-k)$ can be analytically calculated as illustrated in the following example:

Given   :

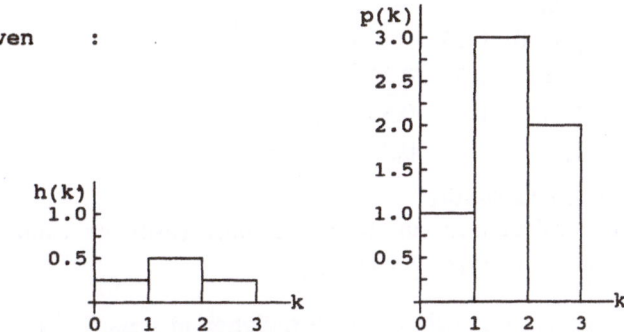

Fig. 3.17 Histograms of system function and system input

Desired: $q(n) = \sum_0^n p(k)h(n-k)$

Solution:

$q(0) = p(0)h(0)$
$q(1) = p(0)h(1) + p(1)h(0)$
$q(2) = p(0)h(2) + p(1)h(1) + p(2)h(0)$
$q(3) = p(0)h(3) + p(1)h(2) + p(2)h(1) + p(3)h(0)$
$q(4) = p(0)h(4) + p(1)h(3) + p(2)h(2) + p(3)h(1) + p(4)h(0)$

$\quad$ ...   ...

$q(0) = 1.0 \cdot 0.25 \qquad\qquad\qquad\qquad\qquad = 0.25$
$q(1) = 1.0 \cdot 0.50 + 3.0 \cdot 0.25 \qquad\qquad = 1.25$
$q(2) = 1.0 \cdot 0.25 + 3.0 \cdot 0.50 + 2.0 \cdot 0.25 \quad = 2.25$
$q(3) = \qquad \div \qquad\quad 3.0 \cdot 0.25 + 2.0 \cdot 0.50 \quad = 1.75$
$q(4) = \qquad \div \qquad\qquad \div \qquad\quad 2.0 \cdot 0.25 \quad = 0.50.$

By use of the symbols for matrices the equations can be written as follows:

$$\begin{bmatrix} q(0) \\ q(1) \\ q(2) \\ q(3) \\ q(4) \end{bmatrix} \overset{>}{=} \overset{>}{q} \quad \begin{bmatrix} 0.25 & \div & \div \\ 0.50 & 0.25 & \div \\ 0.25 & 0.50 & 0.25 \\ \div & 0.25 & 0.50 \\ \div & \div & 0.25 \end{bmatrix} = \bar{h} \quad \begin{bmatrix} 1.00 \\ 3.00 \\ 2.00 \end{bmatrix} \overset{>}{=} \overset{>}{p}$$

$\overset{>}{q} = \bar{h} * \overset{>}{p}.$

Procedure for multiplication of matrices:

$$
*\begin{bmatrix} 1.00 \\ 3.00 \\ 2.00 \end{bmatrix}
$$

$$
\begin{bmatrix}
0.25 & \div & \div \\
0.50 & 0.25 & \div \\
0.25 & 0.50 & 0.25 \\
\div & 0.25 & 0.50 \\
\div & \div & 0.25
\end{bmatrix}
\Rightarrow
\begin{bmatrix}
0.25 \\
1.25 \\
2.25 \\
1.75 \\
0.50
\end{bmatrix}
= \overset{>}{q}
$$

Aid to calculation:
The calculation can be made more easily by using the computer program VAZAO1 (Appendix 9).

Practical Example for the Simulation of Runoff Data

The catchment area under consideration is situated in the tropical climate of Brazil, southwestwards of the capital Joao Pessoa in the Paraiba state. The river in this catchment is called the Rio Mamuaba.

The simulation of discharges firstly serves for verification of the model's reliability, and secondly for estimation of dangerous river conditions caused by extreme precipitation events. The information will be utilized for flood protection connected with a possible reservoir construction that will be necessary to extend the water supply in the capital.

Catchment area: $115.3\,\text{km}^2$ (to Fazenda Mamuaba)

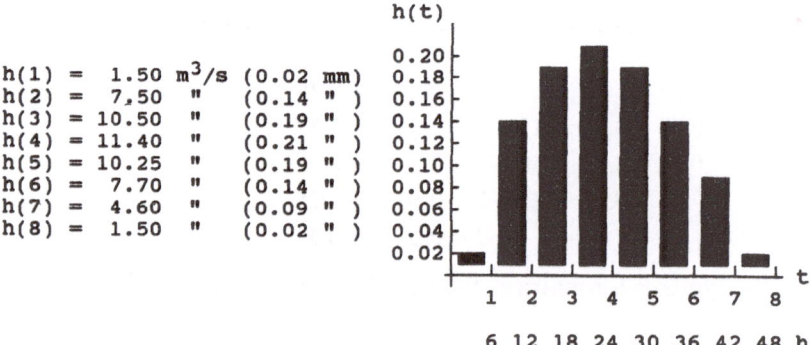

$$
\begin{array}{llll}
h(1) = & 1.50 & \text{m}^3/\text{s} & (0.02 \text{ mm}) \\
h(2) = & 7.50 & " & (0.14 \text{ "}) \\
h(3) = & 10.50 & " & (0.19 \text{ "}) \\
h(4) = & 11.40 & " & (0.21 \text{ "}) \\
h(5) = & 10.25 & " & (0.19 \text{ "}) \\
h(6) = & 7.70 & " & (0.14 \text{ "}) \\
h(7) = & 4.60 & " & (0.09 \text{ "}) \\
h(8) = & 1.50 & " & (0.02 \text{ "})
\end{array}
$$

Fig. 3.18 Unit hydrograph with a time interval of 6 h

To verify the reliability of this model, rainfall data and corresponding runoff data of the period from 20.05.1974 till 24.05.1974 are used.

[$p(t)$ = mean areal precipitation calculated by methods explained above]

p(21 – 03) = 3.68 mm   p(21 – 03) =   –   mm
p(03 – 09) =   –   ”   p(03 – 09) =   –   ”
p(09 – 15) = 15.66 ”   p(09 – 15) = 11.27 ”
p(15 – 21) = 2.04 ”   p(15 – 21) = 6.09 ”

p(21 – 03) =   –   ”   p(21 – 03) = 1.62 ”
p(03 – 09) =   –   ”   p(03 – 09) =   –   ”
p(09 – 15) =   –   ”   p(09 – 15) =   –   ”
p(15 – 21) =   –   ”   p(15 – 21) =   –   ”

Desired: Discharge curve
Solution: convolution of precipitation by the unit hydrograph
Conversion of m^3/s in mm/6 h:

$$1\,m^3/s = 21{,}600 \cdot 1{,}000/(115.3 \cdot 1{,}000{,}000) = 0.1873\,mm/6\,h$$

Aid to calculation:
The calculation can be made more easily by using the computer program
VAZAO1 or VAZAO2 (Appendix 10).

Result:
[q(t) = discharge calculated, q*(t) = discharge measured]

q(21 – 03) = 0.39 m^3/s q*(21 – 03) = 0.30 m^3/s
q(03 – 09) = 2.75 ”   q*(03 – 09) = 3.40 ”
q(09 – 15) = 5.41 ”   q*(09 – 15) = 7.20 ”
q(15 – 21) = 16.05 ”   q*(15 – 21) = 19.60 ”

q(21 – 03) = 21.14 ”   q*(21 – 03) = 20.40 ”
q(03 – 09) = 22.38 ”   q*(03 – 09) = 22.30 ”
q(09 – 15) = 19.94 ”   q*(09 – 15) = 21.80 ”
q(15 – 21) = 14.16 ”   q*(15 – 21) = 15.00 ”

q(21 – 03) = 9.05 ”   q*(21 – 03) = 7.70 ”
q(03 – 09) = 2.65 ”   q*(09 – 15) = 2.20 ”
q(09 – 15) = 1.42 ”   q*(15 – 21) = 5.00 ”
q(15 – 21) = 9.07 ”   q*(03 – 09) = 3.20 ”

q(21 – 03) = 16.16 ”   q*(21 – 03) = 16.00 ”
q(03 – 09) = 20.02 ”   q*(03 – 09) = 21.40 ”
q(09 – 15) = 19.90 ”   q*(09 – 15) = 21.00 ”
q(15 – 21) = 16.41 ”   q*(15 – 21) = 17.80 ”

q(21 – 03) = 11.61 ”   q*(21 – 03) = 11.70 ”
q(03 – 09) = 5.34 ”   q*(03 – 09) = 6.40 ”
q(09 – 15) = 1.43 ”   q*(09 – 15) = 3.40 ”
q(15 – 21) = 0.17 ”   q*(15 – 21) = 1.20 ”

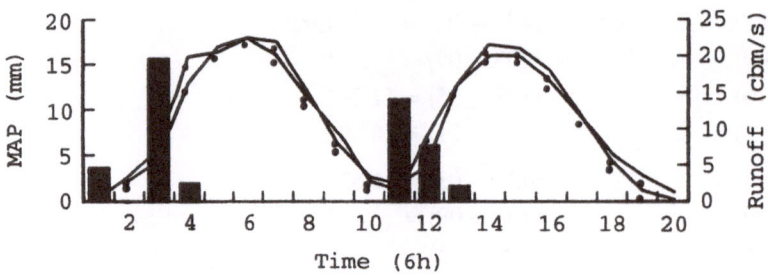

Fig. 3.19 Graphical representation of the results

3.4   Treatment of Variables in the Frequency Domain

3.4.1   Fourier, Laplace and Z-Transformations

Fourier Transformation

The resolution of a time function p(t) [p(t) may be considered a system input function] into series of exponential functions can be written by analogy with Section 3.3.2 in the following form, where the coefficients z(v) are not yet known.

$$
\begin{aligned}
p(t) = z_{-v}\exp(-iw_v t) + \cdots \\
+ z_{-1}\exp(-iwt) + z_0 + z_1\exp(iwt) + \cdots \\
+ z_v\exp(iw_v t)
\end{aligned} \tag{3.73}
$$

or

$$
p(t) = \sum_{n=-\infty}^{n=+\infty} z_n\exp(iw_n t)
$$

with n, v∈N (indices)

$$v \to \infty$$
$$i = \text{complex unit} = \sqrt{-1} \tag{3.74}$$
$$w = \text{angular velocity}$$
$$t = \text{time variable}$$
$$z = \text{complex number}$$

and
$$[(a_n + b_n) + i(b_{-n} - a_{-n})] = z_n$$
$$a, b \in R.$$

Because of the convergence, the resolution of the time function p(t) into series of exponential functions is connected with a set of conditions satisfied by the time function (Dirichlet conditions).

1. The function $p(t)$ has not more than a finite number of maxima and minima in the considered time interval.
2. The function $p(t)$ has not more than a countable number of finite discontinuities in the considered time interval.
3. The function $p(t)$ has not more than a finite number of infinite discontinuities in the considered time interval, and

$$\int_{-T/2}^{+T/2} p(t)\,dt < \infty. \tag{3.75}$$

The unknown coefficients (z_n) in the series of exponentials can be determined by using the orthogonality principle (principle of least squares) as follows:
The resolved time function in the series of exponentials with unknown coefficients in this context is marked by an asterisk.

$$p^*(t, z_n) = \sum_{n=-\infty}^{n=+\infty} \exp(iw_n t), \quad \text{with } -\infty < n < +\infty. \tag{3.76}$$

The error F between the resolved time function $p^*(t, z_n)$ and the given time function $p(t)$ is:

$$F(t, z_n) = [p(t) - p^*(t, z_n)] \text{ and in a quadratic form:}$$
$$F^2(t, z_n) = [p(t) - p^*(t, z_n)]^2. \tag{3.77}$$

The mean squares error E (expected value) is desired to be minimum:

$$E[F^2(t, z_n)] = 1/2v \int_{-T/2}^{+T/2} F^2(t, z_n)\,dt \to \min. \tag{3.78}$$

By this demand we can determine the unknown coefficients the series of exponentials. The mean squares error function has to be differentiated in relation to the coefficients z_j and the derivatives have to be zero.

$$d/dz_j E[F^2(t, z_n)] = 0 \quad \text{with } -\infty < j, n < +\infty \tag{3.79}$$

$$d/dz_j E F^2[(t, z_n)] = d/dz_j \left[1/2v \int_{-T/2}^{+T/2} F^2(t, z_n)\,dt \right] \tag{3.80}$$

$$= 1/2v \int_{-T/2}^{+T/2} d/dz_j F^2(t, z_n)\,dt$$

$$d/dz_j F^2(t, z_n) = d/dz_j [p(t) - p^*(t, z_n)]^2 \tag{3.81}$$

$$d/dz_j F^2(t, z_n) = 2[p(t) - p^*(t, z_n)][-d/dz_j p^*(t, z_n)] \tag{3.82}$$

$$d/dz_j p^*(t, z_n) = d/dz_j \sum_{n=-\infty}^{n=+\infty} z_n \exp(iw_n t) = \exp(iw_j t) \tag{3.83}$$

$$d/dz_j E[F^2(t, z_n)] = 1/v \int_{-T/2}^{+T/2} \left[p(t) - \sum_{n=-\infty}^{n=+\infty} z_n \exp(iw_n t) \right]$$
$$\cdot [-\exp(iw_j t)]\,dt = 0 \tag{3.84}$$

$$\sum_{n=-\infty}^{n=+\infty} z_n \int_{-T/2}^{+T/2} \exp(iw_{n+j}t)\,dt = \int_{-T/2}^{+T/2} p(t)\exp(iw_j t)\,dt. \tag{3.85}$$

We are considering the left member of the equation and insert values for $n+j$:

1. if $n+j=0 \Rightarrow n=-j \Rightarrow z_n = z_{-j}$

$$\int_{-T/2}^{+T/2} dt = T \tag{3.86}$$

2. if $n+j \neq 0$

$$\int_{-T/2}^{+T/2} \exp(iw_{n+j}t)\,dt = [1/iw_{n+j}\exp(iw_{n+j}t)]_{-T/2}^{+T/2} \tag{3.87}$$

$$= 1/iw_{n+j}[\exp(iw_{n+j})T/2 - \exp(-iw_{n+j})T/2]$$
$$= 2/iw_{n+j}\,\text{sen}((w_{n+j})T/2) \quad \text{with } w = 2\pi/T$$
$$= 2/i\dot{w}_{n+j}\,\text{sen}\,[\pi(n+j)] = 0$$

(properties of orthogonal functions).
It follows:

$$\sum_{n=-\infty}^{n=+\infty} z_n \int_{-T/2}^{+T/2} \exp(iw_{n+j}t)\,dt = z_{-j}T = \int_{-T/2}^{+T/2} p(t)\exp(iw_j t)\,dt \tag{3.88}$$

$$= z_n T = \int_{-T/2}^{+T/2} p(t)\exp(-iw_n t)\,dt. \tag{3.89}$$

The equation can be resolved for the coefficients z_n in a considered time interval T:

$$z_n = 1/T \int_{-T/2}^{+T/2} p(t)\exp(-iw_n t)\,dt. \tag{3.90}$$

By substitution of this relation for the unknown coefficients in the equation used for resolution of a given time function in the series of exponentials, we obtain:

$$p(t) = \sum_{n=-\infty}^{n=+\infty} z_n \exp(iw_n t)$$

$$= \sum_{n=-\infty}^{n=+\infty} \left[1/T \int_{-T/2}^{+T/2} p(t)\exp(-iw_n t)\,dt \right] \exp(iw_n t). \tag{3.91}$$

To extend the time interval to infinite ($T \to \infty$), the angular velocity w is involved because of the relation $w = 2\pi/T$. With $T = 2\pi/w$ this yields:

$$p(t) = \sum_{n=-\infty}^{n=+\infty} \exp(iw_n t)\cdot w/2\pi \int_{-\pi/w}^{+\pi/w} p(t)\exp(-iw_n t)\,dt. \tag{3.92}$$

From the limit $w \to dw$ it follows:

$$p(t) = \lim_{w \to 0} \sum_{n=-\infty}^{n=+\infty} \exp(iw_n t) w/2\pi \int_{-\pi/w}^{+\pi/w} p(t) \exp(-iw_n t) dt. \qquad (3.93)$$

Finally, we obtain the following equation:

$$p(t) = 1/2\pi \int_{-\infty}^{+\infty} \exp(iwt) \int_{-\infty}^{+\infty} p(t) \exp(-iwt) dt\, dw, \qquad (3.94)$$

which can be used to resolve a time function $p(t)$ into the series of exponential functions.

This relation is known as the Fourier transformation and can also be written in the following form:

$$
\boxed{
\begin{aligned}
p^{\wedge}(iw) &= \int_{-\infty}^{-\infty} p(t) \exp(-iwt) dt = :\pounds[p(t)] \\[2mm]
p(t) &= 1/2\pi \int_{-\infty}^{+\infty} p^{\wedge}(w) \exp(iwt) dw = :\pounds^{\wedge}[p^{\wedge}(iw)]
\end{aligned}
}
\qquad (3.95)
$$

The function $p^{\wedge}(iw)$ depending on the angular velocity (or frequence) is called the Fourier transform of the time function $p(t)$.

The symbols $\pounds$ and $\pounds^{\wedge}$ stand for transformation and retransformation.

Laplace Transformation

The Laplace transformation is similar to the Fourier transformation. One advantage in using it arises from the fact those functions can also be transformed which do not satisfy the third condition of Dirichlet.

The transformation formulas differ merely by the substitution of complex numbers (s) for purely imaginary numbers (iw). The Laplace transformation is extended for values in the total complex plane.

With $s = a + bi$ and $s \in C$, a, $b \in R$ yields:

$$
\boxed{
\begin{aligned}
p^{\wedge}(s) &= \int_{-\infty}^{-\infty} p(t) \exp(-st) dt = :\pounds[p(t)] \\[2mm]
p(t) &= 1/2\pi i \int_{a-i\infty}^{a+i\infty} p^{\wedge}(s) \exp(st) ds = :\pounds^{\wedge}[p^{\wedge}(s)]
\end{aligned}
}
\qquad (3.96)
$$

The function $p^{\wedge}(s)$ is called the Laplace transform (or image function) of the time function $p(t)$, and it depends on complex numbers.

The symbols $\pounds$ and $\pounds^{\wedge}$ stand for transformation and retransformation.

Z-Transformation

The Z-transformation is frequently used for resolution of discrete time functions. If the argument of time functions is considered as a complex number, the function can be developed into a Laurant series, that is comparable to the Taylor series in the real analysis. Each single term of the series is a complex number that is also an exponential function because of the Euler relation.

Euler relation:

$$\exp(i\alpha) = \cos\alpha + i\operatorname{sen}\alpha \quad \text{or} \quad z = r(\cos\alpha + i\operatorname{sen}\alpha)$$

With $r = 1$ we obtain:

$$\exp(i\alpha) = z \quad \text{or} \quad \exp(i\alpha n) = z^n.$$

Hence, from the Fourier transformation, the Z-transformation is obtained by substitution of complex numbers for the exponentials.

$$
\begin{aligned}
p^\wedge(z) &= \sum_{n=0}^{n=\infty} p(n)z^{-n} =: Z[p(n)] \\
p(n) &= 1/2\pi i \int^c p^\wedge(z)z^{n-1}\,dz =: Z^\wedge[p^\wedge(z)]
\end{aligned}
\tag{3.97}
$$

The complex function $p^\wedge(z)$ is called the Z-transform of the time function $p(n)$. The symbols Z and $Z^\wedge$ stand for transformation and retransformation.

3.4.2   Convolution Theorem

The Laplace transformation can be applied to the convolution integral representing a time function. On being applied in this manner, it will be evident that the Laplace transformation cancels out the procedure of integration, and what remains to be done is only a simple multiplication. In this manner, the calculation of the convolution integral is much simplified.

At first, however, it is necessary to set up Laplace transforms for the functions involved. For elementary functions (impulse function, step function etc.), where time functions have already been resolved, there will be no difficulty.

Proof:

$$q(t) = \int_0^t h(\tau)p(t-\tau)\,d\tau \tag{3.98}$$

$$
\begin{aligned}
\pounds[q(t)] = \pounds\left[\int_0^t h(\tau)p(t-\tau)\,d\tau\right] &= \int_{-\infty}^{+\infty}\left[\int_0^t h(\tau)p(t-\tau)\,d\tau\right]\exp(-st)\,dt \\
&= \int_{-\infty}^{+\infty} p(t-\tau)\exp(-st)\,dt \int_0^t h(\tau)\,d\tau.
\end{aligned}
$$

Using the one sided Laplace transformation, the substitution is: $t - \tau = \sigma$ with $d\sigma/dt = 1$ and für $t \rightarrow \infty$:

$$\pounds[q(t)] = \int_0^\infty p(\sigma)\exp[-s(\sigma + \tau)]\,d\tau \int_0^\infty h(\tau)\,d\tau \qquad (3.99)$$

$$= \int_0^\infty p(\sigma)\exp(-s\sigma)\,d\sigma \int_0^\infty h(\tau)\exp(-s\tau)\,d\tau$$

$$\pounds[q(t)] = \pounds[p(\sigma)]*\pounds[h(\tau)]. \qquad (3.100)$$

By this characteristic of the Laplace transformation for the convolution integral, the calculation of the involved functions becomes considerably simpler.

The symbol ($*$) in this context stands of multiplication.

3.4.3 Synthesis of System Output Functions by Convolution

In Section 3.3.4 the synthesis of the system output functions in the time domain has already been explained, and some examples have been given to illustrate the technique for calculation.

Concerning the frequency domain there is the same aim – synthesis of output functions – but the manner of calculation is different.

The convolution integral in the frequency domain is calculated by the use of the transformation technique.

By applying the convolution theorem there are considerable simplifications. The procedure for calculation can be schematically demonstrated as follows:

Original function   Transform function

$$q(t) = \int_0^t h(\tau)p(t - \tau)\,d\tau \Rightarrow \text{Transformation: } \pounds[p(t)], \pounds[h(\tau)]$$

$$\downarrow$$

Calculation:

$$\pounds[q(t)] = \pounds[h(\tau)]*\pounds[p(\sigma)]$$

$$\downarrow$$

$q(t)$ $\Leftarrow$ Retransformation: $\pounds^\wedge\{\pounds[q(t)]\}$.

Before calculation the Laplace transforms of elementary functions have to be known (impulse function, step function).

Forming the Laplace transform of the step function $\mu(t)$:

$$\pounds[\mu(t)] = \int_0^\infty \mu(t)\exp(-st)\,dt \qquad (3.101)$$

$$= \lim_{x \to \infty} \int_0^x 1 \cdot \exp(-st)\,dt = \lim_{x \to \infty} \left[\exp(-st)\right]/-s \Big|_0^x \tag{3.102}$$

$$= \lim_{x \to \infty} [1 - \exp(-sx)]/s$$

$$= 1/s$$

$$\pounds[a\mu(t)] = \int_0^\infty \mu(t)\exp(-st)\,dt = a/s \quad \text{with } a \in R \tag{3.103}$$

$$\pounds[\mu(t - t_0)] = \int_0^\infty \mu(t - t_0)\exp(-st)\,dt \tag{3.104}$$

$$= \int_0^\infty \mu(x)\exp[-s(x + t_0)]\,dx \quad \text{with } t - t_0 = x$$

$$= \exp(-st_0)\int \mu(x)\exp(-sx)\,dx$$

$$= [\exp(-st_0)]/s.$$

The retransformation yields:

$$\pounds^\wedge(1/s) = \mu(t); \pounds^\wedge(a/s) = a\mu(t); \pounds^\wedge[(\exp(-st_0)/s] = \mu(t - t_0). \tag{3.105}$$

Forming the Laplace transform of the impulse function $\delta(t)$:

$$\pounds[\delta(t)] = \int_0^\infty \delta(t)\exp(-st)\,dt \tag{3.106}$$

$$= \lim_{t \to 0} \int_0^\infty \delta(t)\exp(-st)\,dt = \exp(-0\cdot t)\int_0^\infty \delta(t)\,dt = 1 \tag{3.107}$$

$$\pounds[a\delta(t)] \quad = a\int_0^\infty \delta(t)\exp(-st)\,dt = a \quad \text{with} \quad a \in R \tag{3.108}$$

$$\pounds[\delta(t - t_0)] = \int_0^\infty \delta(t - t_0)\exp(-st)\,dt \tag{3.109}$$

$$= \int_0^\infty \delta(x)\exp[-s(x + t_0)]\,dx \quad \text{with} \quad t - t_0 = x \tag{3.110}$$

$$= \exp(-st_0)\int \delta(x)\exp(-sx)\,dx = \exp(-st_0). \tag{3.111}$$

The retransformation yields:

$$\pounds^\wedge(1) = \delta(t); \pounds^\wedge(a) = a\delta(t); \pounds^\wedge[\exp(-st_0)] = \delta(t - t_0). \tag{3.112}$$

Illustrative example of application to the Laplace transformation in the calculation of the convolution integral:

Given: Time function p(t) as a system imput function resolved in impulse functions.

System reponse function h(t) resolved in impulse functions.

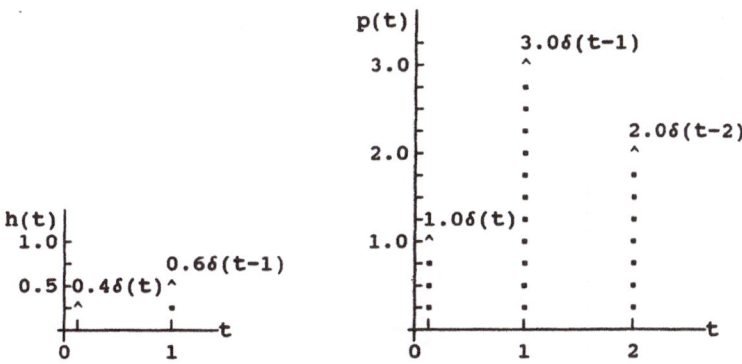

Fig. 3.20 Representation of time functions of the system response, and system input resolved in impulses

Desired: $q(t) = \int\limits_{0}^{t} h(\tau) p(t - \tau) \, d\tau$

Solution:

$$£[h(t)] = \int [0.4\delta(t) + 0.6\delta(t - 1)] \exp(-st) \, dt = 0.4 + 0.6 \exp(-s)$$

$$£[p(t)] = \int [1.0\delta(t) + 3.0\delta(t - 1) + 2.0\delta(t - 2)] \exp(-st) \, dt$$

$$= 1.0 + 3.0 \exp(-s) + 2.0 \exp(-2s)$$

$$£[q(t)] = £[h(t)] * £[p(t)]$$

$$= [0.4 + 0.6 \exp(-s)] \cdot [1.0 + 3.0 \exp(-s) + 2.0 \exp(-2s)]$$

$$= 0.4 + 1.2 \exp(-s) + 0.8 \exp(-2s)$$

$$+ 0.6 \exp(-s) + 1.8 \exp(-2s) + 1.2 \exp(-3s)$$

$$= 0.4 + 1.8 \exp(-s) + 2.6 \exp(-2s) + 1.2 \exp(-3s)$$

$$q(t) = £^{\wedge}\{£[q(t)] = £^{\wedge}(£[h(t)] * £[p(t)]\}$$

$$= 1/2\pi i \int\limits^{c} [0.4 + 1.8 \exp(-s) + 2.6 \exp(-2s)$$

$$+ 1.2 \exp(-3s)] \exp(st) \, ds$$

$$= 0.4\delta(t) + 1.8\delta(t - 1) + 2.6\delta(t - 2) + 1.2\delta(t - 3)$$

q(0) = 0.4
q(1) = 1.8
q(2) = 2.6
q(3) = 1.2

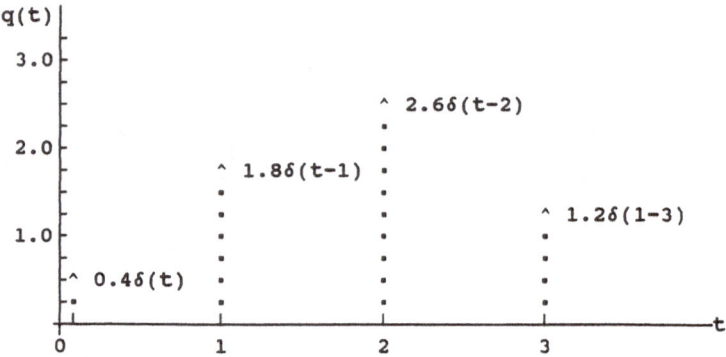

Fig. 3.21 Graphical representation of the result

Aid to calculation:
The calculation can be made more easily by using the computer program VAZAO2 (Appendix 10).

3.5 Particular Linear System (Conceptional Models)

3.5.1 Model of Linear Reservoir

A conceptional model abstracts the rainfall runoff process in a catchment area by a simple physical idea.
It can easily be imagined that in the natural process, the fundamental behaviour is similar to that of the retarding reservoir. The discharge from a catchment caused by rainfall can be estimated by following the pattern of reservoir principles (Hagen-Poiseuille).
For the reservoir model, it can be shown that the system response function is of a specific type because of its characteristic system behaviour or model structure.
The adaptation of the model to the natural process depends on only one parameter of the specific reservoir response function.
On the one hand, if the properties of the natural system have no influence on the type of response function, the reservoir response function is characteristic for the system behaviour, on the other hand, if the modelling of the natural

system depends on the type of response function, the application of the reservoir model will be ineffective although the parameter can be varied.

Therefore, the use of a conceptional model also includes the possibilities of modelling a natural process being confined a priori.

The response function of a linear reservoir is of the following type:

$$h(t) = \alpha \exp(-\alpha t), \quad \alpha \in R. \tag{3.113}$$

The system output function can be calculated by the following convolution integral:

$$q(t) = \int_0^t \alpha \exp[-\alpha(t-\tau)]p(\tau)\,d\tau \tag{3.114}$$

Differentiation of the integral in relation to time t formally yields the differential equation describing the linear reservoir behaviour:

$$d/dt\,q(t) = \int_0^t d/dt\{\alpha \cdot \exp[-\alpha(t-\tau)] \cdot p(\tau)\}\,d\tau \tag{3.115}$$

$$q'(t) = -\int_0^t \alpha^2 \exp[-\alpha(t-\tau)]p(\tau)\,d\tau + \alpha \cdot \exp[-\alpha(t-t)]p(t)$$

$$q'(t) = -\alpha q(t) + \alpha p(t) \tag{3.116}$$

or

$$q'(t) + \alpha q(t) = \alpha p(t). \tag{3.117}$$

An intuitively better derivation of the differential equation describing the linear reservoir behaviour is given in the following illustration:

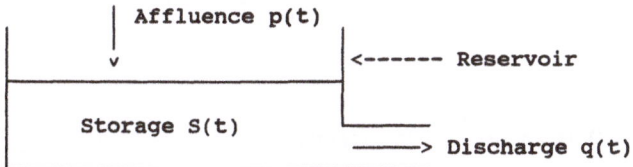

Fig. 3.22 Schematic reservoir system

By applying the principle of continuity we obtain:

$$[p(t) - q(t)]\,dt = dS(t) \tag{3.118}$$

$$p(t) - q(t) = d/dt\,S(t) = S'(t). \tag{3.119}$$

The linearity of the reservoir is given by the assumption that the storage changes proportionally to the discharge.

$$S(t) = k \cdot q(t) \tag{3.120}$$

with k = storage coefficient of the reservoir.

This assumption is a necessary condition for linear reservoir calculation. In the model conception for the rainfall runoff process, it must also be taken into consideration that storage will occur at the same time as discharge occurs. This means that at the same time when the affluence is released storage will be present and accordingly also discharge. Thus, there is no delay between the beginning of affluence and discharge. Consequently, for a linear reservoir model it is impossible to take into consideration any retarding effects of the rainfall runoff process in the catchment area. However, at the start of a rainfall event no change is observed in the discharge because of the translation time of rainwater to the river.

Therefore, the translation process cannot be simulated by the linear reservoir conception, only the retention process by storage can be modelled.

The connection of the storage equation and the equation of continuity provides the differential equation describing the process:

$$p(t) - q(t) = k \cdot d/dt \cdot q(t) = kq'(t) \tag{3.121}$$

$$q'(t) + q(t)/k = p(t)/k \tag{3.122}$$

$$q'(t) + \alpha \cdot q(t) = \alpha \cdot p(t) \quad \text{with} \quad \alpha = 1/k \tag{3.123}$$

Determination of the Reservoir Storage Coefficient

By using a chart of discharge (hydrogram) from a gauge station, the storage coefficient k of a linear reservoir can be determined in the following way:

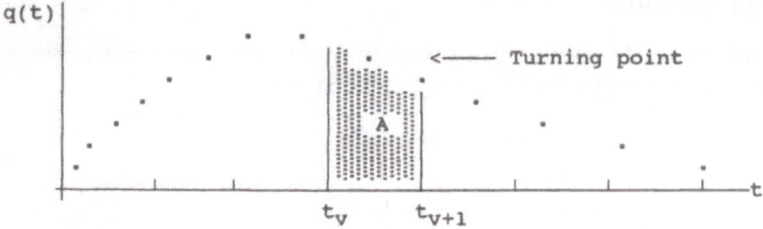

Fig. 3.23 Discharge curve (hydrogram)

$$q(t_v) = \int_0^{t_v} \alpha \exp[-\alpha(t_v - \tau)] p(\tau) \, d\tau \tag{3.124}$$

$$q(t_{v+1}) = \int_0^{t_{v+1}} \exp[-\alpha(t_{v+1} - \tau)] p(\tau) \, d\tau = q(t_v) \cdot \exp(-\alpha) \tag{3.125}$$

$$A = \int_{t_v}^{t_{v+1}} q(t_v) \exp[\alpha(t_v - t)] \, dt = q(t_v) \int_{t_v}^{t_{v+1}} \exp[\alpha(t_v - t)] \, dt \tag{3.126}$$

$$= q(t_v) \left[-1/\alpha \exp[\alpha(t_v - t)] \right]_{t_v}^{t_{v+1}}$$

$$= 1/\alpha [q(t_v) - q(t_v) \exp(-\alpha)]$$

$$= 1/\alpha [q(t_v) - q(t_{v+1})]. \tag{3.127}$$

Finally, we find the following formula for calculation of the storage coefficient:

$$k = 1/\alpha = A/[q(t_v) - q(t_{v+1})].\qquad(3.128)$$

The area A under the discharge curve between the ordinates $q(t_v)$ and $q(t_{v+1})$ below the turning point can be determined in practice using a planimeter. Several discharge curves of different runoff events should be investigated to obtain a representative storage coefficient by averaging.

Example to demonstrate the method:

Given:

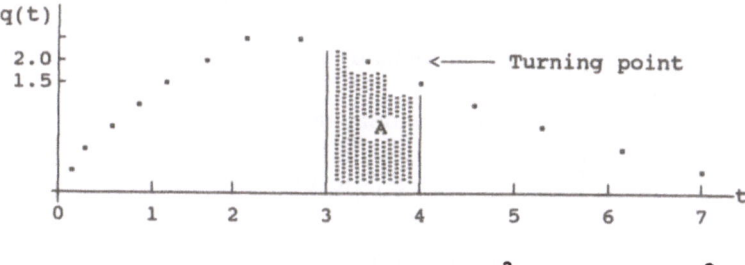

Fig. 3.24 Hydrogram

$$A = 1.75 \ m^3, \quad q(3) = 2.0 \ m^3/h$$
$$q(4) = 1.5 \ m^3/h$$

Desired: k

Solution:

$$k = 1.75/(2.0 - 1.5) = 3.5 \ h.$$

To become aquainted with the technique of linear reservoir modelling it is useful to consider some examples for different types of variables.

Example 1: System input function p(t), continuous

System reponse function h(t), continuous

Given: $k = 2.0 \ h \Rightarrow h(t) = 0.5 \exp(-0.5t)$
$$p(t) = 2t \exp(-t)$$

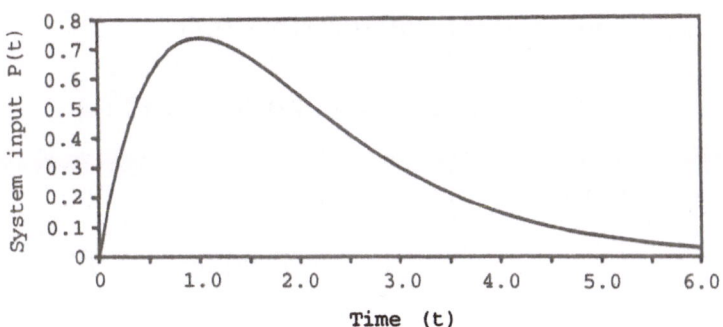

Time (t)

Fig. 3.25 System input function (analytical function)

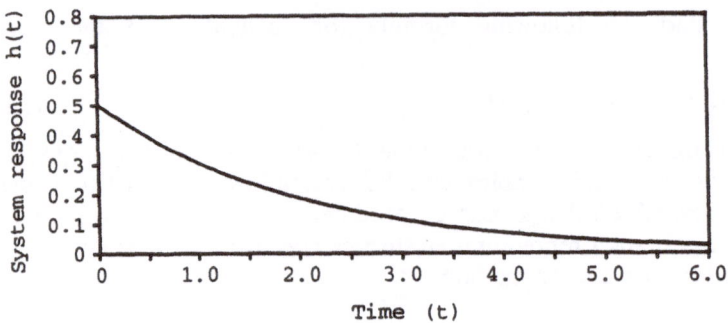

Fig. 3.26 System response function (analytical function)

Desired: System output function q(t)
Solution:
a) Calculation in the time domain:

$$q(t) = \int_0^t \exp[-0.5(t-\tau)]\cdot\tau\cdot\exp(-\tau)d\tau$$

$$= \exp(-t/2)\cdot\int_0^t \tau\cdot\exp(-\tau/2)d\tau \tag{3.129}$$

$$= \exp(-t/2)[4\cdot\exp(-\tau/2)\cdot(-\tau/2-1)]_0^t \tag{3.130}$$

$$= \exp(-t/2)\cdot[4\exp(-t/2)\cdot(-t/2-1)+4] \tag{3.131}$$

$$= -2t\exp(-t)-4\exp(-t)+4\exp(-t/2) \tag{3.132}$$

b) Calculation in the frequency domain:

$$q(t) = \int_0^t \exp[-0.5(t-\tau)]\cdot\tau\cdot\exp(-\tau)d\tau \tag{3.133}$$

Laplace transformation:

$$q(t) = \pounds^\wedge\{\pounds[\exp(-t/2)]*\pounds[\tau\cdot\exp(-\tau)]\} \tag{3.134}$$

$$\pounds(\exp[-t/2]) = \int_0^\infty \exp(-t/2)\cdot\exp(-st)dt \tag{3.135}$$

$$= \int_0^\infty \exp[-(0.5+s)t]dt = 1/(0.5+s) \tag{3.136}$$

$$\pounds[\tau\cdot\exp(-\tau)] = \int_0^\infty \tau\cdot\exp(-\tau)\cdot\exp(-s\tau)d\tau = \int_0^\infty \tau\cdot\exp[-\tau(1+s)]d\tau$$

$$= [[-\tau(1+s)-1]\cdot\exp[-\tau(1+s)]/(1+s)^2]_0^\infty = 1/(1+s)^2 \tag{3.137}$$

$$\pounds[q(t)] = 1/(0.5+s)\cdot 1/(1+s)^2 = 1/[(0.5+s)(1+s)^2] \tag{3.138}$$

$$= A/(0.5+s) + B/(1+s)^2 + C/(1+s). \tag{3.139}$$

Division for partial fractions provides:

$$A = 4; \quad B = -2; \quad C = -4 \tag{3.140}$$

$$\pounds[q(t)] = 4/(0.5 + s) - 2/(1 + s)^2 - 4/(1 + s). \tag{3.141}$$

Retransformation is made by use of a list for Laplace transformation (Schwarz et al. 1965).

$$q(t) = 4\exp(-0.5t) - 2t\exp(-t) - 4\exp(-t) \tag{3.142}$$

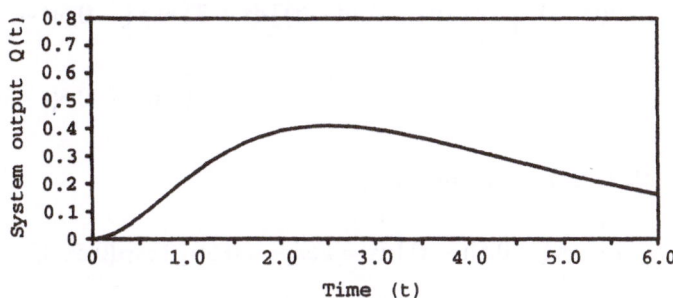

Fig. 3.27 System output function (analytical function)

Example 2: System input function p(t), limited step function
System response function h(t), continuous

Given: $k = 0.5\,h \Rightarrow h(t) = 0.5\exp(-0.5t)$

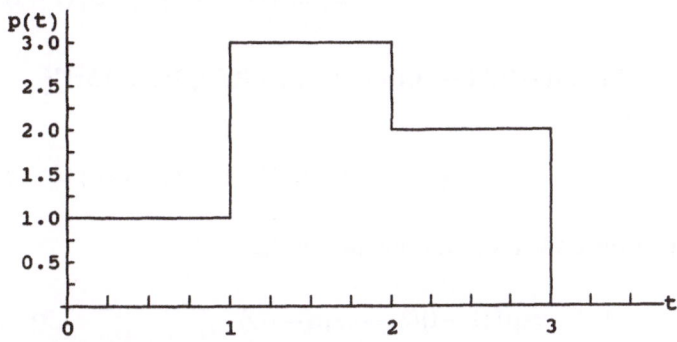

Fig. 3.28 System input function (step function)

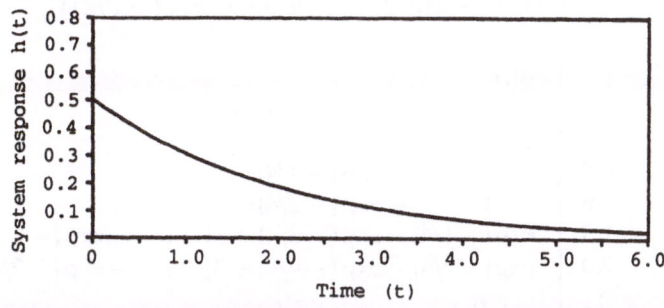

Fig. 3.29 System response function (analytical function)

Desired: System output function q(t)
Solution:
a) Calculation in the time domain:

$$q(t) = \int_0^t 0.5 \exp[-0.5(t-\tau)] \cdot p(\tau) \, d\tau \tag{3.143}$$

$$q(t) = 0.5 \left[1 \int_0^t \exp[-0.5(t-\tau)] \, d\tau + 3 \int_1^2 \exp[-0.5(t-\tau)] \, d\tau \right.$$
$$\left. + 2 \int_2^3 \exp[-0.5(t-\tau)] \, d\tau \right]. \tag{3.144}$$

Calculation of the integrals:

$$1 \int_0^1 \exp[-0.5(t-\tau)] \, d\tau = 1 \exp(-0.5t)[2\exp(0.5\tau)]_0^1$$
$$= 2\{\exp[(1-t)/2] - \exp[-t/2]\} \tag{3.145}$$

$$3 \int_0^2 \exp(-0.5(t-\tau)) \, d\tau = 3 \exp(-0.5t)[2\exp(0.5\tau)]_1^2$$
$$= 6\{\exp[(2-t)/2] - \exp[(1-t)/2]\} \tag{3.146}$$

$$2 \int_0^3 \exp(-0.5(t-\tau)) \, d\tau = 2 \exp(-0.5t)[2\exp(0.5\tau)]_2^3 \tag{3.147}$$

$$= 4\{\exp[(3-t)/2] - \exp[(2-t)/2]\}. \tag{3.148}$$

For the system output function yields:

$$\begin{array}{ll}
q1(t) = \exp[(1-t)/2] - \exp(-t/2) & 0 < t < 1 \\
q2(t) = q1(t) + 3\{\exp[(2-t)/2] - \exp[(1-t)/2]\} & 1 < t < 2 \\
q3(t) = q2(t) + 2\{\exp[(3-t)/2] - \exp[(2-t)/2]\} & 2 < t < 3
\end{array} \tag{3.149}$$

t	q(t)
0.0	0
0.5	$1 \quad -\exp(-1/4)$
1.0	$1 \quad -\exp(-2/4)$
1.5	$\exp(-1/4) - \exp(-3/4) + 3[\ 1 \quad -\exp(-1/4)]$
2.0	$\exp(-2/4) - \exp(-4/4) + 3[\ 1 \quad -\exp(-2/4)]$
..	......

t	q(t)	
0.5	0.22	= 0.22
1.0	0.39	= 0.39
1.5	0.30 + 3·0.22	= 0.96
2.0	0.24 + 3·0.39	= 1.41
2.5	0.19 + 3·0.30 + 2·0.22	= 1.53
3.0	0.14 + 3·0.24 + 2·0.39	= 1.64
3.5	0.11 + 3·0.19 + 2·0.30	= 1.28
4.0	0.09 + 3·0.14 + 2·0.24	= 0.99
4.5	....	= 0.71
5.0	....	= 0.55
5.5	....	= 0.22
6.0	....	= 0.18
..	....	= ..

b) Calculation in the frequency domain:
   Laplace transformation:

$$\mathcal{L}[p(t)] = \int_0^\infty [\mu(t) + 2\mu(t-1) - \mu(t-2) - 2\mu(t-3)]\exp(-st)dt \qquad (3.150)$$

$$= 1/s + (2/s)\exp(-s) - (1/s)\exp(-2s) - (2/s)\exp(-3s) \qquad (3.151)$$

$$\mathcal{L}[h(t)] = \int_0^\infty 1/2 \cdot \exp(-0.5t) \cdot \exp(-st)dt = 0.5(0.5 + s) \qquad (3.152)$$

$$\mathcal{L}[q(t)] = 0.5[1 + 2\exp(-s) - \exp(-2s) - 2\exp(-3s)]/s(0.5 + s)$$

with: $0.5/s(0.5 + s) = 1/s - 1/(0.5 + s)$, is obtained:

$$\mathcal{L}[q(t)] = [1/s - 1/(0.5 + s)] \cdot [1 + 2\exp(-s) - \exp(-2s) - 2\exp(-3s)]$$
$$(3.153)$$

Retransformation:

$$q(t) = [1 - \exp(-0.5t)] + 2\mu(t-1)\{1 - \exp[-0.5(t-1)]\}$$
$$- \mu(t-2)\{1 - \exp[-0.5(t-2)]\}$$
$$- 3\mu(t-3)\{1 - \exp[-0.5(t-3)]\} \qquad (3.154)$$

$q(0,0) = 0$	= 0.00
$q(0.5) = [1 - \exp(-1/4)]$	= 0.22
$q(1.0) = [1 - \exp(-2/4)]$	= 0.39
$q(1.5) = [1 - \exp(-3/4)] + 2[1 - \exp(-1/4)]$	= 0.96
$q(2.0) = [1 - \exp(-4/4)] + 2[1 - \exp(-2/4)]$	= 1.41
$q(2.5) = [1 - \exp(-5/4)] + 2[1 - \exp(-3/4)] - [1 - \exp(-1/4)]$	= 1.53
$q(3.0) = [1 - \exp(-6/4)] + 2[1 - \exp(-4/4)] - [1 - \exp(-2/4)]$	= 1.64
.. = ...	= ...

$$(3.155)$$

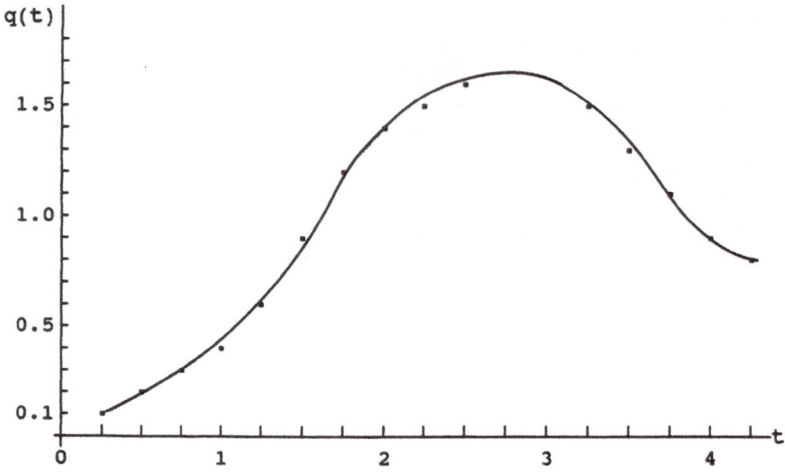

Fig. 3.30 System output function (analytical function)

Example 3: System input function p(t), impulses
           System response function h(t), continuous

Given: $k = 0.5\,h \Rightarrow h(t) = 0.5\exp(-0.5t)$

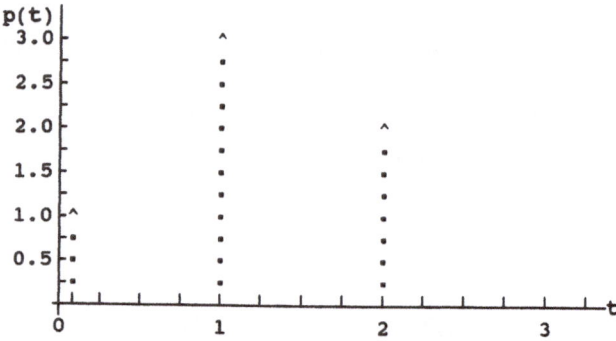

Fig. 3.31 System input function (impulse function)

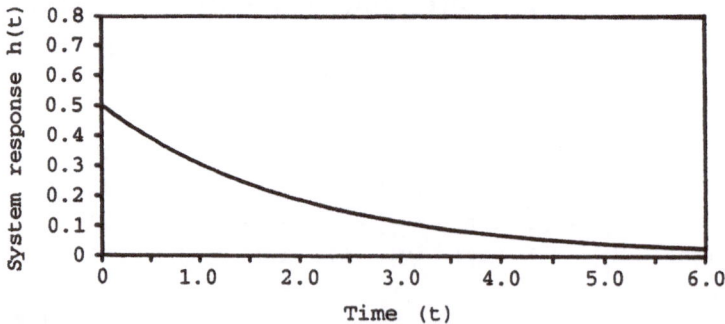

Fig. 3.32 System response function (analytical function)

Desired: System output function q(t)
Solution:
a) Calculation in the time domain:

$$q(t) = \int_0^t 0.5 \exp[-0.5(t-\tau)] \cdot [\delta(\tau) + 3\delta(\tau-1) + 2\delta(\tau-2)] d\tau \qquad (3.156)$$

$$q(t) = 0.5 \int_0^t \exp[-0.5(t-\tau)] \cdot \delta(\tau) d\tau$$

$$+ 3 \int_0^t \exp[-0.5(t-\tau)] \cdot \delta(\tau-1) d\tau$$

$$+ 2 \int_0^t \exp[-0.5(t-\tau)] \cdot \delta(\tau-2) d\tau$$

$$\qquad (3.157)$$

$$q(t) = 0.5 \exp[-t/2] \int_0^t \exp[\tau/2]\delta(\tau) d\tau$$

$$+ 3 \int_0^t \exp[\tau/2]\delta(\tau-1) d\tau$$

$$+ 2 \int_0^t \exp[\tau/2]\delta(\tau-2) d\tau \qquad (3.158)$$

$$
\begin{aligned}
q_1(t) &= 0.5 \exp(-t/2) & &= 0.5 \exp(-t/2),\, 0 < t1 \\
q_2(t) &= 0.5 \exp(-t/2)[1 + 3\exp(0.5)] & &= 3.0 \exp(-t/2),\, 1 < t2 \\
q_3(t) &= 0.5 \exp(-t/2)[1 + 3\exp(0.5) + 2\exp(1)] &= 5.7 \exp(-t/2),\, 2 < t
\end{aligned}
$$

$$\qquad (3.159)$$

t	0.0	0.5	1.0	1.5	2.0	2.5	3.0	3.5	4.0	4.5	5.0
q(t)	0.5	0.4	0.3								
			1.8	1.4	1.1						
					2.1	1.6	1.3	1.0	0.8	0.6	0.5

b) Calculation in the frequency domain:
Laplace transformation:

$$\pounds[p(t)] = \int_0^t [\delta(t) + 3\delta(t-1) + 2\delta(t-2)] \exp(-st) dt \qquad (3.160)$$

$$= 1 + 3\exp(-s) + 2\exp(-2s) \qquad (3.161)$$

$$\pounds[h(t)] = 0.5/(0.5+s)$$

$$\pounds[q(t)] = 0.5[1 + 3\exp(-s) + 2\exp(-2s)]/(0.5+s)$$

$$= 0.5/(0.5+s) + 1.5\exp(-s)/(0.5+s) + \exp(-2s)/(0.5+s).$$

$$\qquad (3.162)$$

Retransformation:

$$q(t) = 0.5 \exp(-0.5t) + 1.5\mu(t-1)\exp[-(t-1)/2]$$
$$+ \mu(t-2)\exp[-(t-2)/2]. \qquad (3.163)$$

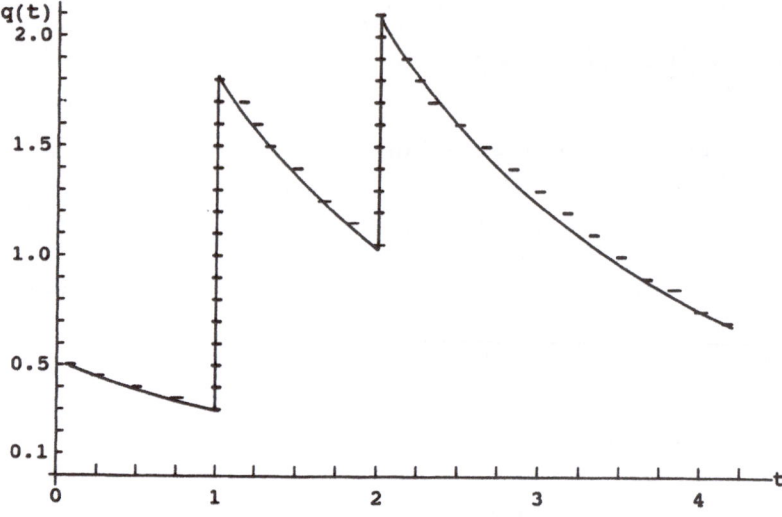

Fig. 3.33 System output function (step function)

Example 4: System input function p(t), impulses
                System response function h(t), impulses

Given: $k = 0.5\,h$

$$h(t) = 0.5 \exp(-0.5t) \cdot [\delta(t) + \delta(t-1) + \delta(t-2) + \cdots]$$

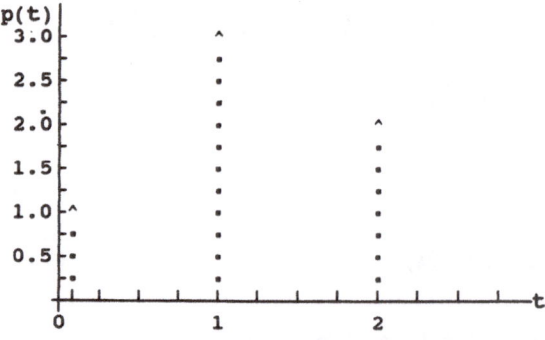

Fig. 3.34 System input function (impulse function)

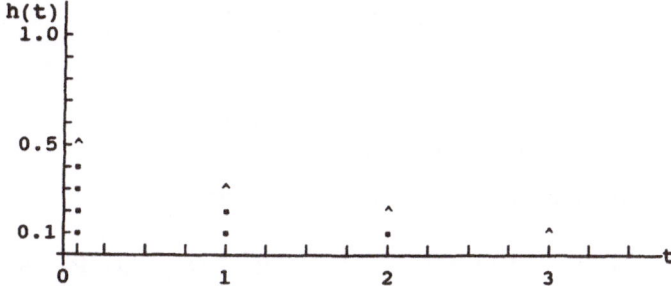

Fig. 3.35 System response function (impulse function)

Desired: System output function q(t)
Solution:
a) Calculation in the time domain:

$$q(t) = \int_0^t 0.5 \exp\left[-(t-\tau)/2\right] \cdot \left[\delta(t-\tau) + \delta(t-1-\tau) + \cdots\right]$$

$$\cdot \left[\delta(\tau) + 3\delta(\tau-1) + 2\delta(\tau-2)\right] d\tau \qquad (3.164)$$

$$q(t) = 0.5 \int_0^t \exp(\tau - t/2) \cdot \left[\delta(t-\tau)\delta(\tau) + 3\delta(t-\tau)\delta(\tau-1) + 2\delta(t-\tau)\delta(\tau-2)\right.$$

$$+ \delta(t-1-\tau)\delta(\tau) + 3\delta(t-1-\tau)\delta(\tau-1)$$

$$\left. + 2\delta(t-1-\tau)\delta(\tau-2) + \cdots\right] d\tau \qquad (3.165)$$

with: $\tau = t,\ \tau = t-1,\ \tau = t-2, \ldots$ is obtained:

$$q(t) = 0.5 \cdot \left[\delta(0)\delta(t-0) + 3\delta(0)\delta(t-1) + 2\delta(0)\delta(t-2)\right.$$

$$+ \delta(-1)\delta(t-0) + 3\delta(-1)\delta(t-1) + 2\delta(-1)\delta(t-2) + \cdots\right]$$

$$+ 0.5 \exp(-1/2) \cdot \left[\delta(1)\delta(t-1) + 3\delta(1)\delta(t-2) + 2\delta(1)\delta(t-3)\right.$$

$$+ \delta(0)\delta(t-1) + 3\delta(0)\delta(t-2) + 2\delta(0)\delta(t-3) + \cdots\right]$$

$$+ 0.5 \exp(-2/2) \cdot \left[\delta(2)\delta(t-2) + 3\delta(2)\delta(t-3) + 2\delta(2)\delta(t-4)\right.$$

$$+ \delta(1)\delta(t-2) + 3\delta(1)\delta(t-3) + 2\delta(1)\delta(t-4)$$

$$\left. + \delta(0)\delta(t-2) + 3\delta(0)\delta(t-3) + 2\delta(0)\delta(t-4) + \cdots\right]$$

$$+ \cdots \qquad (3.166)$$

$$q(t) =$$

$$+ \delta(t) \qquad \cdot\ 0.5$$

$$+ \delta(t-1) \cdot \left[1.5 + 0.5 \exp(-1/2)\right]$$

$$+ \delta(t-2) \cdot \left[1.0 + 1.5 \exp(-1/2) + 0.5 \exp(-2/2)\right]$$

$$+ \delta(t-3) \cdot \left[\quad + 1.0 \exp(-1/2) + 1.5 \exp(-2/2) + 0.5 \exp(-3/2)\right]$$

$$+ \cdots$$

$$(3.167)$$

t	0	1	2	3	4	5	...
q(t)	0.5	1.8	2.0	1.3	0.8	0.5	...

b) Calculation in the frequency domain:
Laplace transformation:

$$\pounds[p(t)] = 1 + 3\exp(-s) + 2\exp(-2s) \tag{3.168}$$

$$\pounds[h(t)] = 0.5 \int_0^\infty [\delta(t) + \delta(t-1) + \delta(t-2) + \cdots] \exp(-t/2)\exp(-st)\,dt \tag{3.169}$$

$$= 0.5\{1 + \exp[-(1/2+s)] + \exp[-2(1/2+s)]$$
$$+ \exp[-3(1/2+s)] + \cdots\} \tag{3.170}$$

$$\pounds[q(t)] =$$
$$+0.5[\quad 1 \quad + \exp(-1/2-s) + \exp(-2/2-2s) + \exp(-3/2-3s) + \cdots]$$
$$+1.5[\exp(-s) + \exp(-1/2-2s) + \exp(-2/2-3s) + \exp(-3/2-4s) + \cdots]$$
$$+1.0[\exp(-2s) + \exp(-1/2-3s) + \exp(-2/2-4s) + \exp(-3/2-5s) + \cdots]. \tag{3.171}$$

Retransformation:

$$q(t) =$$
$$+\delta(t) \qquad \cdot 0.5$$
$$+\delta(t-1)\cdot[1.5 + 0.5\exp(-1/2)]$$
$$+\delta(t-2)\cdot[1.0 + 1.5\exp(-1/2) + 0.5\exp(-2/2)]$$
$$+\delta(t-3)\cdot[\quad + 1.0\exp(-1/2) + 1.5\exp(-2/2) + 0.5\exp(-3/2)]$$
$$+\delta(t-4)\cdot[1.0\exp(-2/2) + 1.5\exp(-3/2) + 0.5\exp(-4/2)]$$
$$+\delta(t-5)\cdot[1.0\exp(-3/2) + 1.5\exp(-4/2) + 0.5\exp(-5/2)]$$
$$+\cdots \qquad \cdot \tag{3.172}$$

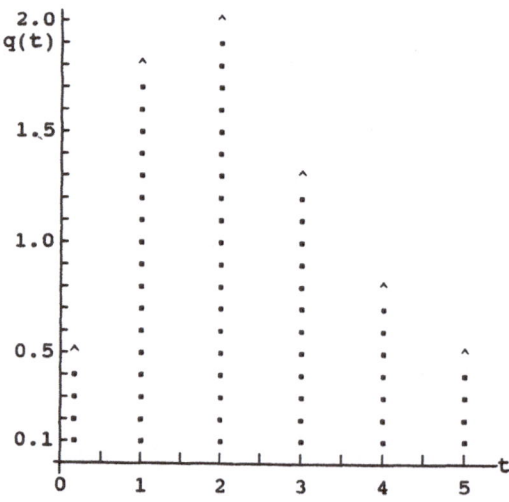

Fig. 3.36 System output function (impulse function)

Aid to calculation:
The calculation can be made more easily by using the computer program
VAZAO3 (Appendix 11).

3.5.2 Model of Linear Cascade

In the linear reservoir model there cannot be taken into consideration any
processes for translation of rainwater through the catchment area. For modelling
the rainfall runoff process including translation facts a linear cascade model is
commonly used. This model, consisting of linear reservoirs ranged side by side
is able to reproduce such effects as occur in the initial phase of the rainfall
runoff process in a catchment where the rainfall began and discharge is not yet
registered.
A linear cascade model of two linear reservoirs can be mathematically
formulated as follows:

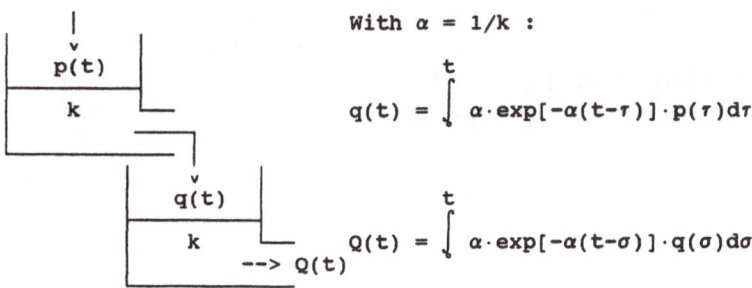

The system response function can be obtained by loading the model system
with the Dirac impulse: $p(t) = \delta(t)$

$$q(t) = \int_0^t \alpha \cdot \exp\left[-\alpha(t - \tau)\right] \cdot \delta(\tau)\, d\tau \tag{3.173}$$

$$h(t) = Q(t) = \int_0^t \alpha \cdot \exp\left[-\alpha(t - \sigma)\right] \cdot q(\sigma)\, d\sigma \tag{3.174}$$

$$\mathcal{L}[q(t)] = \mathcal{L}[\alpha \cdot \exp(-\alpha t)] * \mathcal{L}[\delta(t)] = \alpha/(\alpha + s) \cdot 1 \tag{3.175}$$

$$q(t) = \alpha \cdot \exp(-\alpha t) \tag{3.176}$$

$$\mathcal{L}[h(t)] = \mathcal{L}[\alpha \cdot \exp(-\alpha t)] * \mathcal{L}[\alpha \cdot \exp(-\alpha t)] = [\alpha/(\alpha + s)]^2 \tag{3.177}$$

$$h(t) = \alpha^2 \cdot t \cdot \exp(-\alpha t). \tag{3.178}$$

The system response function provides for $t = 0$ the value $h(0) = 0$, i.e. at the
start of the affluence the discharge is just still zero.

A cascade model consisting of three linear reservoirs can be described as follows:

$$h(t) = \int_0^t \alpha \cdot \exp[-\alpha(t-\tau)] \cdot \alpha^2 \cdot \tau \cdot \exp(-\alpha\tau)\,d\tau \qquad (3.179)$$

$$\pounds[h(t)] = [\alpha/(\alpha+s)] \cdot [\alpha/(\alpha+s)]^2 \qquad (3.180)$$

$$h(t) = (\alpha \cdot \alpha^2/2) \cdot t^2 \cdot \exp(-\alpha t) \qquad (3.181)$$

$$\pounds[t^2 \exp(-\alpha t)] = \int_0^\infty t^2 \exp(-\alpha t)\exp(-st)\,dt = \int_0^\infty t^2 \exp[-(\alpha+s)t]\,dt \qquad (3.182)$$

$$= \{\exp[-(\alpha+s)] \cdot [-(t^2/(\alpha+s))] - [2t/(\alpha+s)^2 - (2/(\alpha+s)(\alpha+s)^2)]\}$$
$$= 2/(\alpha+s)(\alpha+s)^2. \qquad (3.183)$$

For a cascade model with four linear reservoirs is obtained:

$$h(t) = \int_0^t \alpha \cdot \exp[-\alpha(t-\tau)] \cdot \alpha \cdot \alpha^2/2 \cdot \tau^2 \cdot \exp(-\alpha\tau)\,d\tau \qquad (3.184)$$

$$\pounds[h(t)] = [\alpha/(\alpha+s)]^2 \cdot [\alpha/(\alpha+s)]^2$$
$$h(t) = [\alpha^2 \cdot \alpha^2/(1\cdot2\cdot3)] \cdot t \cdot t^2 \cdot \exp(-\alpha t).$$

A cascade model with n linear reservoirs has the following system response function:

$$h(t) = \alpha/(n-1)! \cdot (\alpha t)^{n-1} \cdot \exp(-\alpha t)$$

with $(n-1)! = \Gamma(n)$ is obtained:

$$h(t) = \alpha/[\Gamma(n)] \cdot (\alpha t)^{n-1} \cdot \exp(-\alpha t). \qquad (3.185)$$

This cascade model depends on two parameters n and k res. α. It is well known by the name of the Nash model.

Example for demonstration of the model:
System input function   p(t): impulse function
System output function   h(t): impulse function

Given: $k = 2.0\,h$

$\quad n = 2 \qquad h(t) = \alpha/[\Gamma(n)]\cdot(\alpha t)^{n-1}\cdot\exp(-\alpha t)$

$\quad \Rightarrow h(t) = 0.25\cdot t\cdot\exp(-t/4)\cdot[\delta(t) + \delta(t-1) + \delta(t-2) + \cdots]$

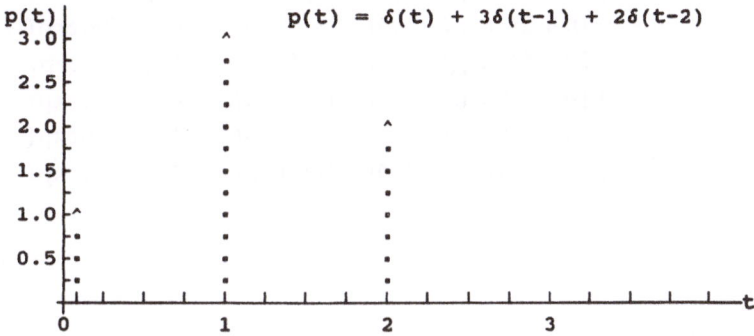

Fig. 3.37 System input function (impulse function)

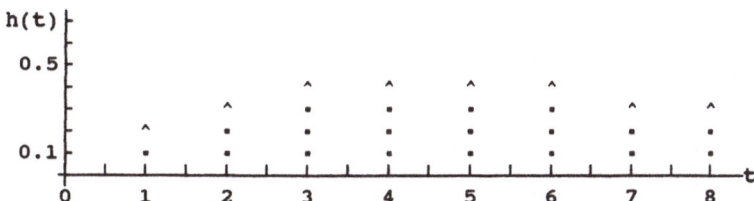

Fig. 3.38 System response function (impulse function)

Desired: System output function q(t)
Solution:
Calculation in the frequency domain:

$$\pounds[p(t)] = 1 + 3\exp(-s) + 2\exp(-2s) \tag{3.186}$$

$$\pounds[h(t)] = 1/4\int_0^\infty [\delta(t) + \delta(t-1) + \delta(t-2) + \cdots]t\exp(-t/4)\exp(-st)\,dt \tag{3.187}$$

$$= 1/4\{0 + \exp[-(1/4 + s)] + 2\exp[-2(1/4 + s)]$$
$$+ 3\exp[-3(1/4 + s)] + \cdots\} \tag{3.188a}$$

$\pounds(q(t)) =$

$\quad + 1/4[0 + 1\exp(-1/4 - 1s) + 2\exp(-2/4 - 2s) + 3\exp(-3/4 - 3s) + \cdots]$

$\quad + 3/4[0 + 1\exp(-1/4 - 2s) + 2\exp(-2/4 - 3s) + 3\exp(-3/4 - 4s) + \cdots]$

$\quad + 2/4[0 + 1\exp(-1/4 - 3s) + 2\exp(-2/4 - 4s) + 3\exp(-3/4 - 5s) + \cdots]$

$$\text{(3.188b)}$$

$q(t) =$

$\quad + \delta(t) \qquad \cdot \; 0$

$\quad + \delta(t-1)\cdot[0 + 1/4\exp(-1/4)]$

$\quad + \delta(t-2)\cdot[0 + 2/4\exp(-2/4) + \quad 3/4\exp(-1/4)]$

$\quad + \delta(t-3)\cdot[0 + 3/4\exp(-3/4) + \quad 6/4\exp(-2/4) + \quad 2/4\exp(-1/4)]$

$\quad + \delta(t-4)\cdot[0 + 4/4\exp(-4/4) + \quad 9/4\exp(-3/4) + \quad 4/4\exp(-2/4)]$

$\quad + \delta(t-5)\cdot[0 + 5/4\exp(-5/4) + 12/4\exp(-4/4) + \quad 6/4\exp(-3/4)]$

$\quad + \delta(t-6)\cdot[0 + 6/4\exp(-6/4) + 15/4\exp(-5/4) + \quad 8/4\exp(-4/4)]$

$\quad + \delta(t-7)\cdot[0 + 7/4\exp(-7/4) + 18/4\exp(-6/4) + 10/4\exp(-5/4)]$

$\cdots \qquad\qquad \cdots\cdots \qquad\qquad\quad \cdots\cdots \qquad\quad \cdots\cdots$

$$\text{(3.189)}$$

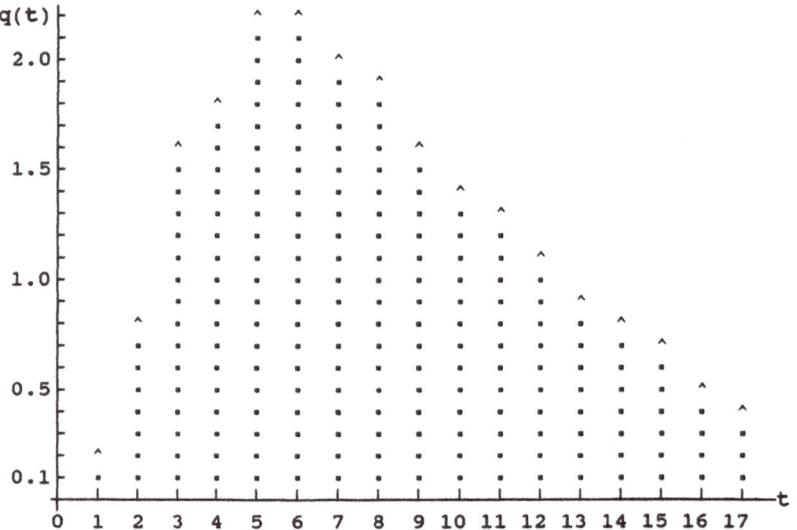

Fig. 3.39 System output function (impulse function)

t	0	1	2	3	4	5	6	7	8	9	10	11	12	13
q(t)	0.0	0.2	0.9	1.6	2.0	2.2	2.2	2.0	1.9	1.6	1.4	1.3	1.1	0.9

Aid to calculation:
The calculation can be made more easily by using the computer program
VAZAO4 (Appendix 12).

3.5.3   Model After Clark

The hydrological rainfall runoff process is influenced by both the storage
capacity and the translation behaviour of the catchment area.
The translation of rainwater through the catchment is carried out by water
particles or drops that move on routes from higher soil levels downwards to
the river. Different soil levels, depending on the relief of the catchment, and the
distance from the strike point of the raindrops to the river, depending on the
channel routes, influence the translation. Together with storage effects, these
factors contribute to the time distribution of river affluence (system output).
The time of translation of rainwater through the catchment can be expressed
as a function of the catchment relief.
For all strike points of rain, single translation times to a certain gauge station
on the river can be estimated by several empirical formulas. A well-known
formula is the following equation of Kirpich:

$$t_c = (0.868 \cdot L^3/H)^{0.385} \quad \text{with } t_c(h), L(km), H(m) \tag{3.190}$$

t_c = translation time
L = distance from strike point of rain to the check point
H = difference of level from strike point of rain to the check point.

Using this formula in the catchment area map, lines can be drawn specified by
a constant translation time to the river gauge station. Such lines are called
isochrones.
Assuming a constant rain distribution over the catchment, the size of areas
between two isochrones represents the amount of rainwater that moves to the
river passing the gauge station. The time sequence of these partial areas forms
a histographical representation of the time development of discharge at the
gauge station caused by rainfall. Such a chart is called a time-area histogram,
where the partial areas are expressed in percent of the total catchment area.
For rainfall of a unit amount causing the discharge by translation, the resulting
time-area histogram represents the system response function.
This model does not take into consideration any storage effects of the catchment
area.
Schematic illustration of a catchment area divided by isochrones and the
resulting time-area histogram are demonstrated as follows:

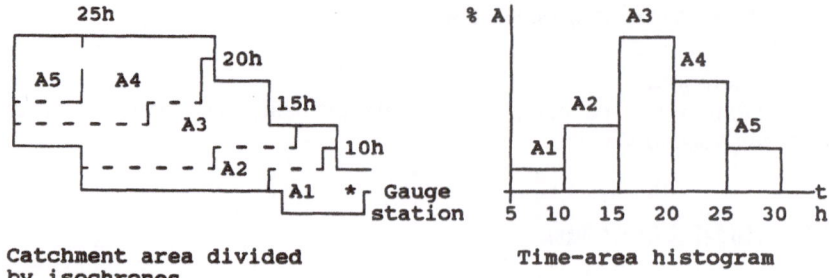

Catchment area divided
by isochrones

Time-area histogram

Fig. 3.40 Catchment with isochrones and time-area graph

Example to demonstrate the isochrone model:
Given:

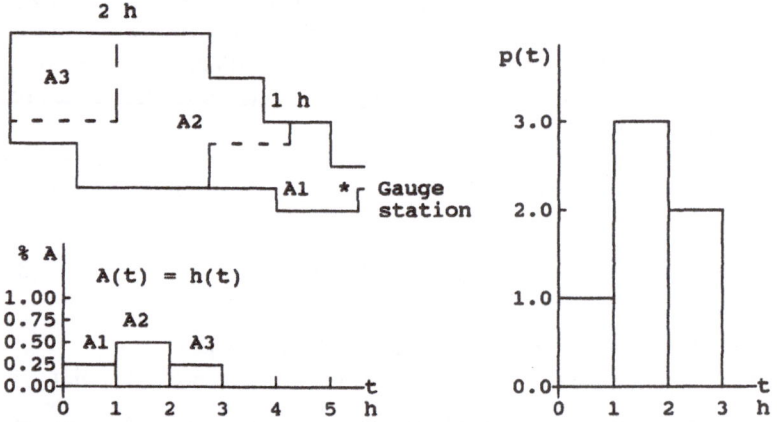

Fig. 3.41 Catchment with isochrones, time-area graph, and system input function

Desired: $q(t) = \sum_0^t p(k)h(t-k)$

Solution:

```
p(0) = 1·0.25                          = 0.25
p(1) = 1·0.50 + 3·0.25                 = 1.25
p(2) = 1·0.25 + 3·0.50 + 2·0.25        = 2.25
p(3) =           3·0.25 + 2·0.25       = 1.75
p(4) =                    2·0.25       = 0.50
```

Using the isochrone pattern to model the rainfall runoff process the property of storage in the catchment area will not be included.

To compensate for this disadvantage in the model, a good solution is to combine the isochrone model with the linear reservoir model, which is able to represent only storage properties. Such a new model, in which the linear reservoir submodel is placed behind the isochrone submodel, is known by the name of Clark-model (Clark 1945)

For calculation of the isochrones in the isochrone submodel a grid net of a chosen size can be used. The translation time from each grid point in the centre of the square to the system outlet is calculated by use of the Kirpich formula. The points with equal translation times in a chosen time interval sequence are connected by a straight line, whereby the isochrones are depicted.

A rainfall runoff model consisting of two submodels a linear reservoir, and an isochrone model constructed in such a way is known by the name of the HYREUN model (Hydrological Research Unit) (Schultz 1968).

3.6   Identification of Linear Systems

In Sections 3.3.4 and 3.4.3 we were concerned with the problem of synthesis of system output functions. The system input function p(t) and the system response function h(t) are known. Using the convolution integral, a calculation algorithm has been given to determine the system output function q(t).

The difficulty in linear system analysis and its application to hydrological systems consists in the determination of the system response function h(t).

The definition of the system response function represents a discharge curve at a gauge station caused by a rainfall unit impulse. Obviously, such a curve cannot be obtained by measurement because in nature there is no rainfall in the form of an impulse.

If the rainfall distribution over a watershed is assumed to be constant at a longer time interval than an instant, the produced discharge curve registered at a gauge station can be used to determine the response function of the system. All ordinates of that curve have to be divided by a constant factor to obtain the system response. Unfortunately, the assumption of a constant rainfall distribution in nature cannot be justified.

There is thus no alternative but to resolve the convolution integral in relation to the response function h(t) for known functions of the system input p(t) and of the system output q(t).

The system output function is obtained from discharge registration charts of a gauge station on the river. The system input function can be determined by using registration charts from the rain gauge stations in the catchment area.

These functions are empirical can so far not free from bias. Mathematically considered, it would be appropriate to use the solution of the Volterra integral equation of the first kind with the constraints given in Section 3.3.3.

3.6.1   Method of Differentiation

The Volterra integral equation of the first kind can be resolved by the differentiation method if the function $p(t, \tau)$, the kernel function, is a polynomial function.

To utilize this method for solving the convolution integral equation, the empirical function $p(t)$ representing the areal precipitation in the rainfall runoff model has to be substituted for an analytical function of the polynomial type. The technique consists of adapting a polynomial function of a chosen degree to the empirical function, whereby the mean squares error is minimum.

Moreover, the empirical discharge function $q(t)$ has to be substituted for an analytical function.

With a good adapted analytical polynomial function for the areal precipitation $p(t)$ and an analytical one for the discharge $q(t)$, the solution of the convolution integral for the system response $h(t)$ can be formulated as follows:

$$q(t) = \int_0^t p(t - \tau) h(\tau) \, d\tau \tag{3.191}$$

$$d/dt \, q(t) = d/dt \int_0^t p(t - \tau) h(\tau) \, d\tau \tag{3.192}$$

$$= \int_0^t d/dt \, p(t - \tau) h(\tau) \, d\tau + p(t - t) h(t). \tag{3.193}$$

Example for demonstrating this method:
Given: $p(t) = t \cdot \exp(-t)$

$$q(t) = t^2 \cdot \exp(-t); \quad q(t) = \int_0^t p(t - \tau) h(\tau) \, d\tau$$

Desired: $h(t)$
Solution:

$$d/dt \, q(t) = (2t - t^2) \cdot \exp(-t) \tag{3.194}$$

$$d/dt \int_0^t (t - \tau) \exp[-(t - \tau)] h(\tau) \, d\tau \tag{3.195}$$

$$= \int_0^t d/dt (t - \tau) \exp[-(t - \tau)] h(\tau) \, d\tau + 0 \tag{3.196}$$

$$= \int_0^t \exp[-(t - \tau)][-(t - \tau) + 1] h(\tau) \, d\tau \tag{3.197}$$

$$= \int_0^t \exp[-(t - \tau)] h(\tau) \, d\tau - \int_0^t (t - \tau) \exp[-(t - \tau)] h(t) \, d\tau \tag{3.198}$$

$$= \int_0^t \exp[-(t - \tau)] h(\tau) \, d\tau - q(t). \tag{3.199}$$

A partial solution is:

$$\int_0^t \exp[-(t-\tau)]\,h(\tau)\,d\tau = q(t) + d/dt\,q(t) = 2t\cdot\exp(-t). \qquad (3.200)$$

By differentiating again we obtain:

$$d/dt[2t\cdot\exp(-t)] = (2-2t)\exp(-t) \qquad (3.201)$$

$$d/dt\int_0^t \exp[-(t-\tau)]\,h(\tau)\,d\tau = \int_0^t d/dt\,\exp[-(t-\tau)]\,h(\tau)\,d\tau + 1\cdot h(t) \qquad (3.202)$$

$$= -\int_0^t \exp[-(t-\tau)]\,h(\tau)\,d\tau + h(t) \qquad (3.203)$$

$$= -2t\cdot\exp(-t) + h(t). \qquad (3.204)$$

We obtain finally:

$$(2-2t)\exp(-t) = -2t\cdot\exp(-t) + h(t) \qquad (3.205)$$

$$h(t) = 2\cdot\exp(-t). \qquad (3.206)$$

Two special cases of the kernal function are considered:
1. Special case: $p(t) = \mu(t)$

$$d/dt\,q(t) = d/dt\int_0^t \mu(t-\tau)\,h(\tau)\,d\tau = \int_0^t \delta(t-\tau)\,h(\tau)\,d\tau = h(t) \qquad (3.207)$$

$$d/dt\,q(t) = h(t). \qquad (3.208)$$

If the time distribution of the effective rain for a catchment area can be considered to be of uniform intensity over a long period of time (generally not given in nature, but auxiliarily assumed), and the subsequent discharge curve is available, the system response function can be obtained by differentiation.

Such a system response function is also called instantaneous Unit Hydrograph (IUH). It represents a system response caused by a rainfall step function.

In the discretized variables, the shape of the system response function depends on the size of the chosen interval of discretization.

Depending on the size of the catchment area, it is advisable to use the following time intervals for discretization (Nelson de Sousa Pinto 1973).

Catchment area (km^2)	Time interval for discretization (h)
> 2600	12, 24
260 to 2600	6, 8 or 12
50	2

In practice, often the problem arises how to use a known system response function for different discretized time intervals. The transformation of a system response function from one given discrete time intervals into another can be made by use of the S-curve construction.

The S-curve represents a system response function caused by a constant rainfall input function of 1.0 mm for each time interval of unit 1 and an infinite duration.

For a desired time interval of discretization, the S-curve is displaced about this interval on the time axis, and the differences or ordinates between this new S-curve and the old one are divided by the value of the new time interval. By this procedure we obtain a system response function related to the desired time interval.

If the time interval of the desired discretization is infinitesimal, the procedure with the S-curve is equivalent to the differentiation. The result is the system response function called instantaneous unit hydrograph, mentioned above.

An illustrative example is given to demonstrate the procedure:

Given:

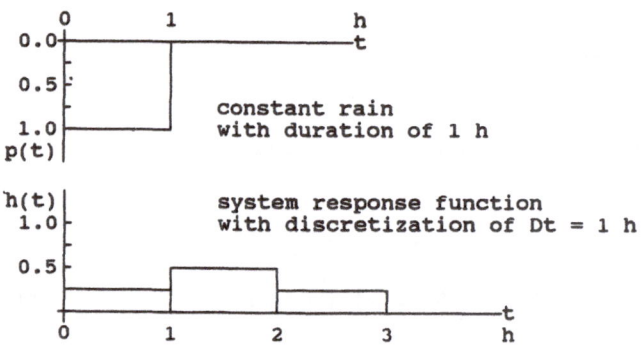

Fig. 3.42 System response function discretized for 1 h

Desired:

1. System response function related to a discretization of Dt = 1/2 h
2. System response function as an instantaneous unit hydrograph

Solution:

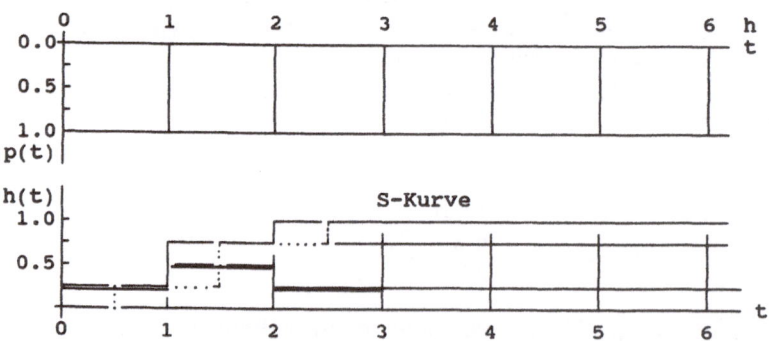

Fig. 3.43 S-curve construction

1. The system response function related to the discretization of Dt = 1/2 h is
   obtained by the differences of ordinates between the 1/2 h displaced S-curve
   and the old one divided by 1/2.

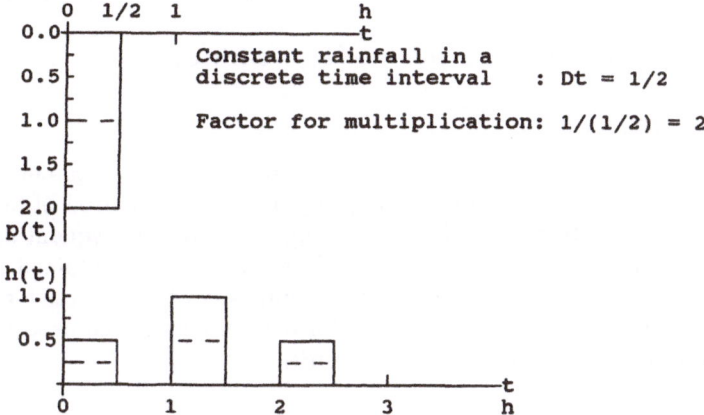

Fig. 3.44 Construction of system response discretized for 1/2 h

or

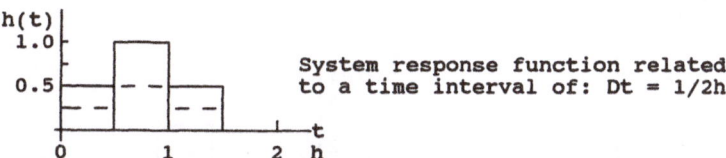

Fig. 3.45 System response function discretized for 1/2 h

2. The system response function related to a infinitesimal time interval is
   obtained by differentiation of the S-curve, whereas the system input function
   approximates to an impulse.

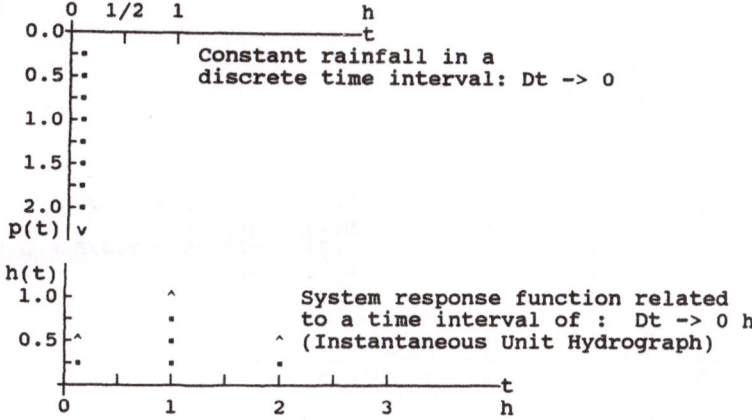

Fig. 3.46 Instantaneous unit hydrograph as an impulse function

2. Special Case: $p(t) = \alpha\delta d(t), a\in R$

$$q(t) = \int_0^t \alpha\delta(t - \tau)\, h(\tau)\, d\tau = \alpha \int_0^t \delta(t - \tau)\, h(\tau)\, d\tau = \alpha h(t) \tag{3.209}$$

$$h(t) = q(t)/\alpha. \tag{3.210}$$

If the time distribution of the effective rain for a catchment area can be considered to be a uniform intensity over a short period of time (generally not given in nature, auxiliarily assumed), and if the subsequent discharge curve is available, the system response function can be obtained by dividing the discharge ordinates by the uniform intensity of the effective rain.

An illustrative example is given for demonstrating the procedure:

Given: $p(t) = \alpha\delta(t)$,   with $\alpha = 5; q(t)$

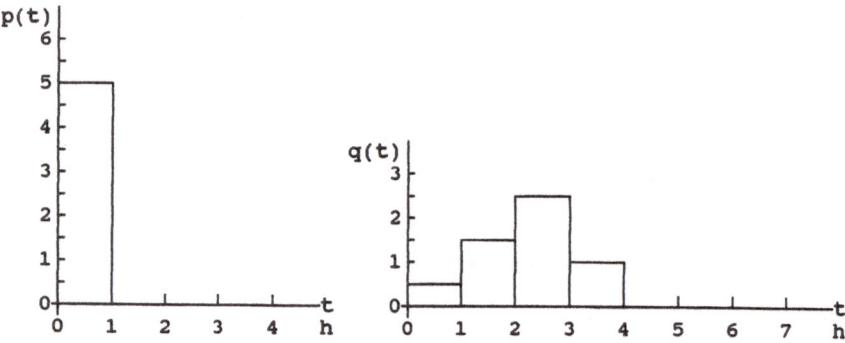

Fig. 3.47 Histograms of system input and system output

Desired: $h(t)$

Solution: $h(t) = q(t)/5$

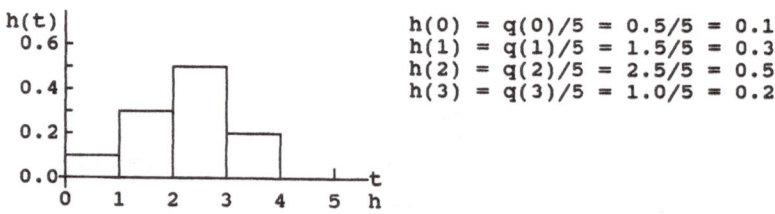

$h(0) = q(0)/5 = 0.5/5 = 0.1$
$h(1) = q(1)/5 = 1.5/5 = 0.3$
$h(2) = q(2)/5 = 2.5/5 = 0.5$
$h(3) = q(3)/5 = 1.0/5 = 0.2$

Fig. 3.48 Histogram of the system response

3.6.2 Method of Transform Technique

The following method is based on the convolution theorem treated in Section 3.4.2, and it will be explained for different transformation techniques.

Laplace Transformation
In the frequency domain the convolution is represented by multiplication of the transforms of the involved functions.

$$\pounds[q(t)] = \pounds[p(\sigma)] * \pounds[h(\tau)]. \tag{3.211}$$

This equation can easily be solved for the transform of the unknown system response function by dividing the transform of the known input function.

$$\pounds[h(\tau)] = \pounds[q(t)]/\pounds[p(t)]. \tag{3.212}$$

By retransformation there is obtained the system response in the time domain.

$$h(\tau) = \pounds^{\wedge}\{\pounds[q(t)]/\pounds[p(\sigma)]\}. \tag{3.213}$$

On the condition that the transforms and retransforms of the involved functions exist and can easily be calculated, the transformation technique will be advantageously applied to solve integral equations.
An illustrative example is given to demonstrate the method:

Given: System output function q(t)
       System input function p(t)

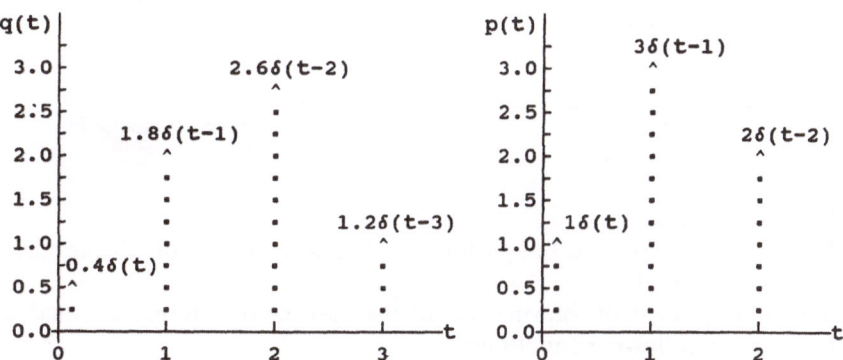

Fig. 3.49 Impulse functions of system input and output

Desired: System response function h(t)
Solution:
Laplace transformation

$$\pounds[q(t)] = \int_0^\infty [0.4\delta(t) + 1.8\delta(t-1) + 2.6\delta(t-2) + 1.2\delta(t-3)] \exp(-st)\,dt$$

$$= 0.4 + 1.8\exp(-s) + 2.6\exp(-2s) + 1.2\exp(-3s) \qquad (3.214)$$

$$\pounds[p(t)] = \int_0^\infty [1.0\delta(t) + 3.0\delta(t-1) + 2.0\delta(t-2)] \exp(-st)\,dt$$

$$= 1.0 + 3.0\exp(-s) + 2.0\exp(-2s). \qquad (3.215)$$

Division

$$\pounds[h(t)] = \qquad\qquad\qquad\qquad\qquad\qquad\qquad\qquad (3.216)$$

$$[0.4 + 1.8\exp(-s) + 2.6\exp(-2s) + 1.2\exp(-3s)]/$$
$$[1 + 3\exp(-s) + 2\exp(-2s)] = 0.4 + 0.6\exp(-s)$$

Retransformation

$$h(t) = \pounds^\wedge \{\pounds[h(t)]\}$$

$$= 1/2\pi i \int_c [0.4 + 0.6\exp(-st)]\,ds = 0.4\delta(t) + 0.6\delta(t-1) \qquad (3.217)$$

Result

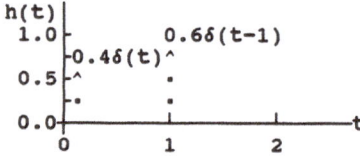

Fig. 3.50 System response (impulse function)

Z-Transformation

In the transformation technique for discrete time functions the Z-transformation is preferably used.
Substituting the Z-transformation for the Laplace transformation analogously leads to the following equations:

$$Z[q(t)] = Z[p(\sigma)] * Z[h(\tau)] \qquad (3.218)$$

$$Z[h(\tau)] = Z[q(t)]/Z[p(\sigma)] \qquad (3.219)$$

$$h(\tau) = Z^\wedge \{Z[q(t)]/Z[p(\sigma)]\}. \qquad (3.220)$$

Some Z-transforms of commonly used time functions are derived as follows. A more extended summary can be found in the literature (Schwarz and Friedland 1965).

1. $f(n) = \alpha\delta(n-k)$ $\qquad\qquad\qquad$ (3.221)

$\qquad Z[f(n)] = \sum_0^\infty \alpha\delta(n-k)z^{-n} = \alpha z^{-k}$ $\qquad$ (3.222)

2. $f(n) = a^n$ $\qquad\qquad\qquad\qquad\qquad$ (3.223)

$\qquad Z[f(n)] = \sum_0^\infty a^n \cdot z^{-n} = \sum_0^\infty (a/z)^n = 1/(1-a/z)$ $\qquad$ (3.224)

3. $Z[f(n)] = 1/(z-a) = 1/(1-a/z)z = 1/(1-a/z)a - z/(az)$

$$= [1/(1-a/z) - z^0]/a \qquad (3.225)$$

$\qquad f(n) = [a^n - \delta(o)]/a = (a^n - 1)/a$ $\qquad$ (3.226)

The following illustrative example may be used to become familiar with this method.

Given: system input function p(t)
$\qquad\qquad$ system output functions q(t)

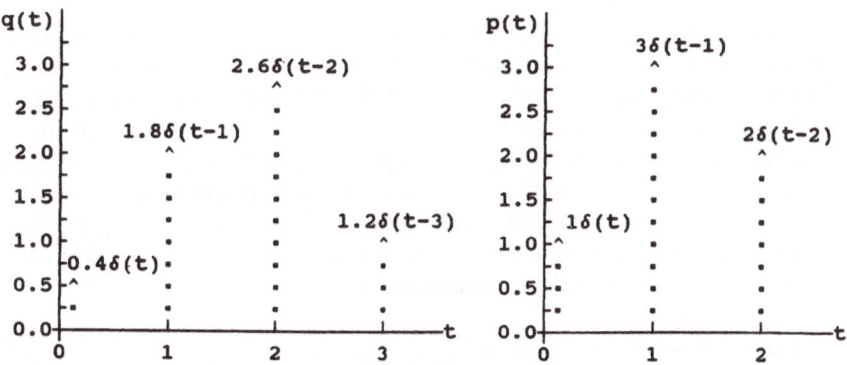

Fig. 3.51 System input and output functions (impulses)

Desired: system response function h(t)
Solution:

$$Z[q(t)] = \sum_0^\infty [0.4\delta(t) + 1.8\delta(t-1) + 2.6\delta(t-2) + 1.2\delta(t-3)]z^{-t} \qquad (3.227)$$

$$= 0.4 + 1.8z^{-1} + 2.6z^{-2} + 1.2z^{-3} \qquad (3.228)$$

$$Z[p(t)] = \sum_0^\infty [1.0\delta(t) + 3.0\delta(t-1) + 2.0\delta(t-2)]z^{-t} \qquad (3.229)$$

$$= 1.0 + 3.0z^{-1} + 2.0z^{-2} \qquad (3.230)$$

$$Z[h(t)] = (0.4 + 1.8z^{-1} + 2.6z^{-2} + 1.2z^{-3})/(1 + 3z^{-1} + 2z^{-2}) \qquad (3.231)$$

$$= (0.4 + 1.8z^{-1} + 2.6z^{-2} + 1.2z^{-3}):(1 + 3z^{-1} + 2z^{-2}) \qquad (3.232)$$

$$= 0.4 + 0.6z^{-1}. \qquad (3.233)$$

Result

$$h(t) = Z^{\wedge}\{Z[h(t)]\} = 0.4\delta(t) + 0.6\delta(t-1) \qquad (3.234)$$

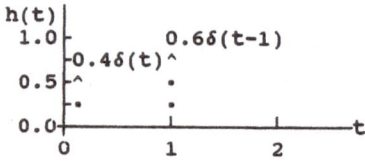

Fig. 3.52 Impulses of the system response

Aid to calculation:
The calculation can be made more easily by using the computer program IDENT1 (Appendix 13).

Consideration of Stability
To calculate the system response function, the application of transformation methods allows a simplification of the procedure for the present.
If the input and output time functions of the system are discretized, it is always possible to obtain a resolution in elementary functions, e.g. δ-functions.
In the frequency domain, the system response function is given by the quotient obtained in dividing the input function by the output function. Concerning the Z-transformation, the input and output functions are represented by polynominals in z, consequently the system response is calculated by dividing a numerator polynomial by a denominator one.

$$h(z) = \frac{q_0 z^0 + q_1 z^{-1} + q_2 z^{-2} + \cdots + q_k z^{-k} + \cdots + q_{n-1} z^{n-1} + q_n z^n}{p_0 z^0 + p_1 z^{-1} + p_2 z^{-2} + \cdots + p_k z^{-k}} = \frac{N(z)}{D(z)}$$

$$(3.235)$$

with $0 \leq k < n$ and p = Ordinate of effective rain per time interval
q = Ordinate of surface runoff per time interval

For causal hydrological systems, the duration of rain over the catchment area is always shorter than the duration of the occurrence of produced surface runoff recorded by a curve or histogram at the river gauge station. Consequently, the order of polynomial must be so that: $0 \leq k < n$.
The difference $n - k$ represents the time interval between the end of rain and the end of surface runoff. This period, called recession, is that of the falling portion of the surface runoff curve (hydrograph).
For a real hydrological system, the input function is a bounded function, as is the output function also. Mathematically it can be formulated as follows:

For $|p(t)| < N$ the system produces $|q(t)| < M$, with $M, N \in R$.

Systems with such properties are called stable systems.

The system response function of a hydrological system is equal to the kernel function in the convolution integral equation. Solving the integral equation for a stable hydrological system by transform methods, the kernel function is obtained by retransformation with the following formula:

$$h(n) = Z^{\wedge}\{Z[h(t)]\} = 1/2\pi i \int_C^{} q^{\wedge}(z)/p^{\wedge}(z)z^n \, dz, \tag{3.236}$$

where C is a circle of radius r enclosing all the singularities of the function $q^{\wedge}(z)/p^{\wedge}(z)z^n$.

Stable hydrological systems are characterized by the following constraints:

1. The singularities or poles of the function $q^{\wedge}(z)/p^{\wedge}(z)z^n$ lie within the unit circle centred on the origin of the z-plane.
   (For retransformation only it is sufficient that the poles lie within the radius of convergence but for the stability of the system the radius of convergence has to be one because the series Σa^n only converges for a <1, see Sect. 3.6.2 Z-transformation, 2).
2. The ordinates of the kernel function (system response) are positive, $h(t) \geqq 0$ for all t.
   (According to the hydrological definition of the unit hydrograph any impulse of effective rain can only produce a runoff recorded by a curve with positive ordinates. Based on a complex or multi-period rainfall event, of course, it is possible to model a runoff curve by a kernel function with negative and positive ordinates. However, in the response function of any hydrological system, there are never negative ordinates).

Because of these constraints on the linear system, the time functions of the effective rain and of the surface runoff are connected in a specific restrictive manner.

Multiplying the quotient h(z) with z and rewriting as a proper fraction yields:

$$h(z) = q_0/p_0 + \frac{a_1 z^{n-1} + a_2 z^{n-2} + a_3 z^{n-3} \cdots + a_k z^{n-k} \cdots + q_{n-1}z + q_n}{p_0 z^n + p_1 z^{n-1} + p_2 z^{n-2} \cdots + p_3 z^{n-k}}, \tag{3.237}$$

with $a_k = q_k - q_0 \cdot p_k/p_0$.

The coefficients of the numerator and the denominator polynominal represent the ordinates of the surface runoff function and of the function of effective rain. Therefore, these coefficients are positive real numbers.

By decomposition of the denominator in a product of polynominal zeros there is obtained:

$$h(z) = q_0/p_0 + \frac{a_1 z^{n-1} + a_2 z^{n-2} + a_3 z^{n-3} \cdots + a_k z^{n-k} \cdots + q_{n-1}z + q_n}{(z-z_0)(z-z_1)(z-z_2)(z-z_3)\cdots(z-z_k)p_0 z^{n-k}}. \tag{3.238}$$

The zero-crossings of the numerator and of the denominator polynomial are either real numbers or conjugate complex numbers. The rewriting in partial fractions yields:

$$h(z) = [q_0 + A_0/(z - z_0) + A_1/(z - z_1) + \cdots + A_k/(z - z_k)$$
$$+ B_1/z \qquad + B_2/z^2 \qquad + \cdots + Bn - k/z^{n-k}]/p_0 \qquad (3.239)$$

By retransformation is obtained:

$$h(n) = [Ao(z_0^n - 1)/z_0 + A_1(z_1^n - 1)/z_1 + \cdots + A_k(z_k^n - 1)/zk$$
$$+ q_0\delta(n) + B_1\delta(n - 1) + B_2\delta(n - 2) + \cdots + B_{n-k}\delta(k)]/p_0. \qquad (3.240)$$

To satisfy the constraints mentioned above, particularly if $p(n)$, $q(n)$, and $h(n) \geq 0$, the zero-crossings of the denominator different from zero must be equal to the zero-crossings of the numerator different from zero. Only in this case do the coefficients $A_0, \ldots, A_k$ vanish and it remains a kernel function as a sum of δ-functions with positive real coefficients.

If the zero-crossings of the denominator different from zero are not equal to the zero-crossings of the numerator different from zero, the kernel function shows one negative value at least (oscillates), whereby the constraints for stability are not maintained.

This can be proved by the following consideration.

Zero-crossings of the denominator:

a) real negative zero-crossings $A_k/(z + z_0)p_0$
   The retransformation formula yields: $A_k[(-z_0)^n - 1]/p_0 z_0$, and for n as an odd number negative values will be obtained in the system response function.
b) real positive zero-crossings $A_k/(z - z_0)p_0$
   By multiplication of the products of polynomial zeros $(z - z_0)(z - z_1) \cdots (z - z_k)$ negative values will be obtained in the denominator polynome.
c) conjugate complex zero-crossings $A_k/[z - (a + bi)][-(a - bi)]p_0$
   For $a < 0$ it is referred to a) and for $a > 0$ it is referred to b).

Therefore, for stable hydrological systems it is necessary that in the transforms of the functions of effective rain and surface runoff, the zero-crossings different from zero are equal to one another. This is important when considering the rainfall runoff behaviour of a catchment because only particular functions of rainfall and runoff specified by the criteria of stability are suitable for modelling. The strategy in the unit hydrograph approach could be carried out so that the surface runoff curve is assumed to be reliable and the corresponding effective rain function is determined by methods including both the principle of continuity and the criteria of stability.

For an areal rainfall distribution the ordinates have to be changed in such a way that for the transformed polynominal of rainfall the zero-crossings different from zero are equal to those of the transformed polynomial of the surface runoff, whereby the condition of continuity has been maintained. In this way, in the Φ-index or percentage loss rates methods, the conditions for stability have to be taken into account. However, in general, the solution of the response function, i.e. the quotient of the transformed polynomials, will not be unique. If in the numerator polynominal there are more zero-crossings than in the denominator, and as only the number of the zero-crossings is known, but not which of that number will be equalized by the changing procedure, help in deciding which zero-crossings are to be selected is given by additional hydrological specifications.

The method is demonstrated on the basis of practical data taken from Nelson de Sousa Pinto (1973).

Given: Catchment area, Rio Capivari in Parana (Brazil)
       Area: 1058 km^2 (gauge station: Praia Grande)
       Discretized time interval: 1 day
       Rainfall event and discharge corresponding
       Gauge stations: A = Estacao Experimental
                      B = Estacao Bocaiuva
                      C = Estacao Praia Grande

The areal precipitation was calculated by the method of the arithmetic mean. The calculation for effective rain was made by the method of percentage loss rates.

Time	Discharge		Precipitation			Areal P.	Eff. R.
			A	B	C		
–	m^3/s	mm	mm	mm	mm	mm	mm
6.2.1954	3.3	0.27	11.0	–	21.3	8.64	1.36
7.2.1954	14.6	1.20	33.8	47.0	46.8	44.16	6.98
8.2.1954	48.4	3.95	25.4	37.2	1.5	24.12	3.81
9.2.1954	46.2	3.77	–	–	–	–	–
10.2.1954	27.6	2.26	–	–	–	–	–
11.2.1954	8.8	0.72	–	–	–	–	–
	148.9	12.17				76.92	12.15

Conversion from m^3 to mm: $f = 60 \cdot 60 \cdot 24/(1058 \cdot 1000) = 0.08166$
Coefficient of discharge: Psi $= 12.15/76.92 = 0.158$
$$\Rightarrow \text{loss rate: } 84.2\%$$
Desired: System response function (unit hydrograph) h(n)
Solution:
In the Z-domain the system response by convolution is given by the following quotient:

$$h(z) = \frac{0.27z^5 + 1.20z^4 + 3.95z^3 + 3.77z^2 + 2.26z + 0.72}{1.26z^5 + 6.98z^4 + 3.81z^3} \qquad (3.241)$$

$$h(z) = \frac{0.27(z + 0.60)(z^2 + 3.34z + 10.21)(z^2 + 0.50z + 0.43)}{(z + 4.50)(z + 0.62)z^3} . \qquad (3.242)$$

The zero-crossings different from zero of the denominator polynomial are not equal to those of the numerator polynominal. The model is not stable because oscillations arise in the system response.

1. Change of zero-crossings of the denominator polynomial:

$$N^*(z) = a(z^2 + 3.34z + 10.21) \quad \text{with } a = \Sigma ai / \Sigma a{*}i$$
$$= 12.15/14.55 = 0.835 \tag{3.243}$$

$$N^*(z) = 0.835z^2 + 2.79z + 8.52. \tag{3.244}$$

Loss rates per day: $\text{Psi}(1) = 0.835/8.64 = 0.09 \quad (91\%)$
$\text{Psi}(2) = 2.79/44.16 = 0.06 \quad (94\%)$
$\text{Psi}(2) = 8.52/24.12 = 0.35 \quad (65\%)$.

System response by the first changing of zero-crossings:

$$h1^*(z) = 0.27(z + 0.60)(z^2 + 0.5z + 0.43)/0.835$$
$$0.32z^3 + 0.35z^2 + 0.23z + 0.10 \tag{3.245}$$

$$h1^*(n) = 0.32\delta(n) + 0.35\delta(n-1) + 0.23\delta(n-2) + 0.10\delta(n-3). \tag{3.246}$$

2. Change of zero-crossings of the denominator polynomial:

$$N^*(z) = a(z^2 + 0.50z + 0.43) \quad \text{with } a = \Sigma ai / \Sigma a{*}i$$
$$= 12.15/1.93 = 6.29 \tag{3.247}$$

$$N^*(z) = 6.29z^2 + 3.15z + 2.70. \tag{3.248}$$

Loss rates per day: $\text{Psi}(1) = 6.29/8.64 = 0.73 \quad (27\%)$
$\text{Psi}(2) = 3.15/44.16 = 0.07 \quad (93\%)$
$\text{Psi}(2) = 2.70/24.12 = 0.11 \quad (89\%)$.

System response by the second changing of zero-crossings:

$$h2^*(z) = 0.27(z + 0.60)(z^2 + 3.3z + 10.21)/6.29 \tag{3.249}$$
$$= 0.04z^3 + 0.17z^2 + 0.52z + 0.26 \tag{3.250}$$

$$h2^*(n) = 0.04\delta(n) + 0.17\delta(n-1) + 0.52\delta(n-2) + 0.26\delta(n-3).$$

The effective rain distribution determined by the first changing of zero-crossings is given preference when calculating the system response function, because the tendency of the percentage loss rates to decrease with time, and an increase in discharge caused by saturation of the soil-water capacity (influence of antecedent precipitation) seems to be hydrologically justified.

However, we cannot totally leave out of account that, during the event, the meteorological conditions (wind, temperature) possibly shift, so that an increase in evaporation and loss rates follows.

In this case the second changing of zero-crossings should be preferred.

If the constraints for stability are not maintained, and there are minor errors in the input-output data, a problem will arise in determining the impulse response function of a linear hydrological system because an oscillatory function will be obtained. In this case stability can be reached additionally by curve-smoothing. For this technique see Neumann and Marsily (1976), Blank et al. (1971) and Bruen et al. (1984).

3.6.3   Method of Matrix Approach

In discretized variables, the convolution integral represents the sum of products of functions. The sum can be written in the form of matrices:

$$q(n) = \sum_{0}^{n} p(n-k)h(k) \tag{3.251}$$

$$\overset{>}{q} = \overset{>}{p} * \overset{>}{h} \tag{3.252}$$

and in more detail:

$$\overset{>}{q} = \begin{bmatrix} q(0) \\ q(1) \\ q(2) \\ \cdot \\ n+m-1 \end{bmatrix} \quad \bar{p} = \begin{bmatrix} p(0) & \cdot & \cdot & n \\ p(1) & p(0) & \cdot & \cdot \\ p(2) & p(1) & p(0) & \cdot \\ \cdot & \cdot & \cdot & \cdot \\ m & \cdot & \cdot & \cdot \end{bmatrix} \quad \overset{>}{h} = \begin{bmatrix} h(0) \\ h(1) \\ h(2) \\ \cdot \\ n \end{bmatrix} \tag{3.253}$$

The matrix equation represents a system of the following linear equations:

$$
\begin{aligned}
q(0) &= h(0)p(0) &&\to n \\
q(1) &= h(0)p(1) + h(1)p(0) \\
q(2) &= h(0)p(2) + h(1)p(1) + h(2)p(0) \\
q(3) &= \cdot\cdot \quad\downarrow\quad + h(1)p(2) + h(2)p(1) + \cdot\cdot \\
q(4) &= \cdot\cdot \quad\downarrow\quad + \cdot\cdot \quad + h(2)p(2) + \cdot\cdot \\
\cdot\cdot &= \cdot\cdot \quad m \quad + \cdot\cdot \quad + \cdot\cdot \quad + \cdot\cdot \\
n+m-1 &
\end{aligned} \tag{3.254}
$$

To resolve this system of linear equations for (n) unknown quantities, there are $(n+m-1)$ equations available. Therefore, the system of linear equations is $(m-1)$-fold over-determined.

The resolvability of such a system of linear equations depends on the rank of the matrix and of the extended matrix of the coefficients. The matrix is of the rank (r) if is there at least one subdeterminant of (r) columns different from zero, and all subdeterminants of $(r+1)$ columns vanish.

The present matrix of coefficients has (n) columns and (m) rows, and with $p(0), \ldots, p(m) \neq 0$ and $m > n$ it follows that the rank is $r = n$.

The extended matrix of coefficients consists of the p-matrix, and the column q-vector is added on its right side. This matrix has $(n+1)$ columns and (m) rows, whereby its rank is $r = n+1$.

The condition for resolvability of the system is as follows:

All the subdeterminants of the extended matrix of coefficients with columns $(n+1)$ must vanish.

(In other words, the rank of the extended matrix of coefficients has to be equal to the number of unknown quantities)

In practice, this condition cannot commonly be maintained because the coefficients of the matrix are represented by measurements subject to errors. Otherwise, the condition of linearity is merely a non-proved assumption on which the matrix equation is based. Therefore, the system of linear equations cannot commonly be resolved so as to provide exact practical results.

A solution to these difficulties consists in searching for an approximation that allows the resolvability of the system with a minimum of errors. The results will then still be useful for practice.

In this approximation the system of linear equations is changed in such a way that the matrix of coefficients is able to provide a unique solution with minimized error. For calculation the least squares method is used.

The result by approximation is noted by the vector $\vec{H}$, and the corresponding vector for error is noted as: $\vec{e}$.

The vector equation for the error yields:

$$\vec{q} - (\bar{p} * \vec{H}) = \vec{e} \quad \text{with} \quad \vec{e} \to 0 \text{ follows: } \vec{H} \to \vec{h}.$$

The least squares method requires: $E(\vec{e}^2) \Rightarrow \min.$

There follows the condition: (E = expected value, mean)

$$E[(\bar{p} * \vec{H} - \vec{q})^2] = (\bar{p} * \vec{H} - \vec{q})^T * (\bar{p} * \vec{H} - \vec{q}) = 0 \tag{3.255}$$

$$(\vec{H}^T * \bar{p}^T - \vec{q}^T) * (\bar{p} * \vec{H} - \vec{q}) = 0 \tag{3.256}$$

$$(\vec{H}^T * \bar{p}^T) * (\bar{p} * \vec{H}) - (\vec{H}^T * \bar{p}^T) * \vec{q} - \vec{q}^T * (\bar{p} * \vec{H}) + \vec{q}^T * \vec{q} = 0$$

Reducing to higher terms with $\bar{p} * (\bar{p}^T * \bar{p})^{-1} * \bar{p}$ yields:

$$(\vec{H}^T * \bar{p}^T * \bar{p} - \vec{q}^T * \bar{p}) * (\bar{p}^T * \bar{p})^{-1} * (\bar{p}^T * \bar{p} * \vec{H} - \bar{p}^T * \vec{q}) = 0 \tag{3.257}$$

Because $(\bar{p}^T * \bar{p})^{-1}$ is not equal to zero, it must be:

$$\bar{p}^T * \bar{p} * \vec{H} - \bar{p}^T * \vec{q} = 0, \tag{3.258}$$

to minimize the error.

This equation can be resolved for the vector H of approximation.

$$\vec{H} = (\bar{p}^T * \bar{p})^{-1} * \bar{p}^T * \vec{q}. \tag{3.259}$$

In this equation it is evident that the multiplication of the p-matrix and its transposed one shows a new quadratic matrix that makes the original system of the linear equations resolvable. The coefficients of the original matrix are changed, whereby the caused error is minimal.

The result thus obtained does not necessarily satisfy the constraints for stability of the hydrological systems. It is therefore possible that the obtained system response function oscillates. In this case the function has to be stabilized, and methods of smoothing can be used as is shown in the literature (Neumann and Marsily 1976; Blank et al. 1971; Bruen et al. 1984; Kavvas 1973).

An example illustrating the method is given as follows:

Given:

$$\vec{q} = \begin{bmatrix} 0.4 \\ 1.8 \\ 2.6 \\ 1.2 \end{bmatrix} \quad \vec{p} = \begin{bmatrix} 1 \\ 3 \\ 2 \end{bmatrix}. \tag{3.260}$$

Desired: h(n)
Solution:

$$h(n) \approx [H(n)] = \vec{H} = (\bar{p}^T * \bar{p})^{-1} * \bar{p}^T * \vec{q} \tag{3.261}$$

$$* \begin{bmatrix} 1 \\ 3 \quad 1 \\ 2 \quad 3 \\ 2 \end{bmatrix} = \bar{p}$$

$$\bar{p}^T = \begin{bmatrix} 1 & 3 & 2 \\ 1 & 3 & 2 \end{bmatrix} = \begin{bmatrix} 14 & 9 \\ 9 & 14 \end{bmatrix} = (\bar{p}^T * \bar{p}) \tag{3.262}$$

$$* \begin{bmatrix} a11 & a12 \\ a21 & a22 \end{bmatrix} = (\bar{p}^T * \bar{p})^{-1} \quad \begin{matrix} 14a11 + 9a21 = 1 & a11 = 0.1217 \\ 14a12 + 9a22 = 0 & a12 = -0.0783 \\ 9a11 + 14a21 = 0 & a21 = -0.0783 \\ 9a12 + 14a22 = 1 & a22 = 0.1217 \end{matrix} \tag{3.263}$$

$$\begin{bmatrix} 14 & 9 \\ 9 & 14 \end{bmatrix} = \begin{bmatrix} 1 & 0 \\ 0 & 1 \end{bmatrix}$$

$$* \begin{bmatrix} 1 & 3 & 2 \\ 1 & 3 & 2 \end{bmatrix}$$

$$\begin{bmatrix} 0.1217 & -0.0783 \\ -0.0783 & 0.1217 \end{bmatrix} = \begin{bmatrix} 0.1217 & 0.2868 & 0.0087 & -0.1565 \\ -0.0783 & -0.1131 & 0.2086 & 0.2434 \end{bmatrix}$$

$$= (\bar{p}^T * \bar{p})^{-1} * \bar{p}^T \tag{3.264}$$

$$\ast \begin{bmatrix} 0.4 \\ 1.8 \\ 2.6 \\ 1.2 \end{bmatrix}$$

$$\begin{bmatrix} 0.1217 & 0.2868 & 0.0087 & -0.1565 \\ -0.0783 & -0.1131 & 0.2086 & 0.2434 \end{bmatrix} = \begin{bmatrix} 0.3997 \\ 0.5995 \end{bmatrix}$$

$$= (\bar{p}^T \ast \bar{p})^{-1} \ast \bar{p}^T \ast \overset{>}{q} = \overset{>}{H}. \tag{3.265}$$

This example is a specific case because the subdeterminants of $(n+1)$ vanish, and the exact solution for the system response is of $h(0) = 0.4$ and $h(1) = 0.6$. By this method an obviously good approximation is obtained.
Determination of the rank of matrices:
(a) matrix of coefficients (b) extended matrix of coefficients

$$A = \begin{bmatrix} p(0) & 0 \\ p(1) & p(0) \\ p(2) & p(1) \\ 0 & p(2) \end{bmatrix} \qquad B = \begin{bmatrix} p(0) & 0 & q(0) \\ p(1) & p(0) & q(1) \\ p(2) & p(1) & q(2) \\ 0 & p(2) & q(3) \end{bmatrix}. \tag{3.266}$$

Subdeterminants $r = 2$

$$D2 = \begin{vmatrix} p(0) & 0 \\ p(1) & p(0) \end{vmatrix} \neq 0 \quad D2 = \begin{vmatrix} p(0) & 0 \\ p(1) & p(0) \end{vmatrix} \neq 0. \tag{3.267}$$

Subdeterminants $r = 3$ (by marginal completion)

$$D3 = \begin{vmatrix} p(0) & 0 & q(0) \\ p(1) & p(0) & q(1) \\ p(2) & p(1) & q(2) \end{vmatrix} \to 0$$

$$D3' = \begin{vmatrix} p(0) & 0 & q(0) \\ p(1) & p(0) & q(1) \\ 0 & p(2) & q(3) \end{vmatrix} \to 0$$

$$D3 = p(0)[p(0)q(2) - p(1)q(1)] + q(0)[p(1)p(1) - p(2)p(0)]$$
$$= 1(1 \cdot 2.6 - 3 \cdot 1.8) + 0.4(3 \cdot 3 - 2 \cdot 1) = 0 \tag{3.268}$$

$$D3' = p(0)[p(0)q(3) - p(2)q(1)] + q(0)[p(1)p(2)] \tag{3.269}$$
$$= 1(1 \cdot 1.2 - 2 \cdot 1.8) + 0.4(3 \cdot 2) = 0. \tag{3.270}$$

In another example with data from field measurements the usefulness of the method is demonstrated.
(It is interesting to compare the following result with that obtained by the Z-transformation method.)
The data are taken from Nelson da Sousa Pinto (1973).

Given: Catchment area, Rio Capivari in Parana (Brazil)
      Area: 1058 km^2 (gauge station: Praia Grande)
      Discretized time interval: 1 day
      Rainfall event and discharge corresponding
      Gauge stations: A = Estacao Experimental, B = Estacao Bocaiuva,
                C = Estacao Praia Grande.

The areal precipitation was calculated by the method of the arithmetic mean. The calculation for effective rain has been made by the method of percentage loss rates.

Date	Discharge		Precipitation			Areal P.	Eff. R.
			A	B	C		
–	m^3/s	mm	mm	mm	mm	mm	mm
6.2.1954	3.3	0.27	11.0	–	21.3	8.64	1.36
7.2.1954	14.6	1.20	33.8	47.0	46.8	44.16	6.98
8.2.1954	48.4	3.95	25.4	37.2	1.5	24.12	3.81
9.2.1954	46.2	3.77	–	–	–	–	–
10.2.1954	27.6	2.26	–	–	–	–	–
11.2.1954	8.8	0.72	–	–	–	–	–
	148.9	12.17				76.92	12.15

Conversion from m^3 to mm: $f = 60 \cdot 60 \cdot 24/(1058 \cdot 1000) = 0.08166$
Coefficient of discharge: Psi = $12.15/76.92 = 0.158$
$$\Rightarrow \text{loss rate: } 84.2\%.$$

Desired: System response function (unit hydrograph) h(n)
Solution:

p(0) = 1.36 mm	q(0) = 0.27 mm	h(0) = 0.09 mm
p(1) = 6.98 ”	q(1) = 1.20 ”	h(1) = 0.47 ”
p(2) = 3.81 ”	q(2) = 3.95 ”	h(2) = 0.25 ”
p(3) = –	q(3) = 3.77 ”	h(3) = 0.19 ”
p(4) = –	q(4) = 2.26 ”	h(4) = –
p(5) = –	q(5) = 0.72 ”	h(5) = –.

The calculation was made by use of the computer program IDENT2.
Aid to calculation:
The calculation can be made more easily by using the computer program IDENT2 (Appendix 14).

3.7 Linear Systems with Stochastic Inputs

In a hydrological system the time functions of precipitation and of corresponding discharge are formed by measurements. They are empirical and not analytical functions. Therefore, the measured values are not free from bias, and the empirical functions are frequently contaminated by noise. This is a property of physical phenomena with a stochastic variety of functions that can only be described statistically.

This behaviour of empirical functions gathered under a certain aspect is shown by a random process.

For a mathematical description of hydrological systems it is essential to determine the relationships of the statistical properties between the involved random processes.

Statistical properties can be presented by deterministic time functions, like the average behaviour or the correlation function, and they have a considerable influence on the system description.

3.7.1   Random Processes and Linear Systems

In the literature, linear systems are frequently denoted by linear filters, where the statistical properties of the random process at the output are determined in terms of the known statistical properties of the input process. In this chapter, when discussing some rainfall runoff models, it is convenient to use this term. Before starting the discussion of some linear filters, it will be explained what is a random process, and in which way we can find out its statistical properties. The term of random process $[X(x, t)]$ stands for the statistical information about a group or ensemble of the random functions. Random functions are commonly time functions, where its ordinates are governed by a probability distribution. A single random function is denoted by a sample function.

Concerning a stationary hydrological system the effective rain distribution can be considered a random process $[P(p, t)]$ at the input, and the runoff distribution at a river gauge station represents a random process $[Q(q, t)]$ at the output.

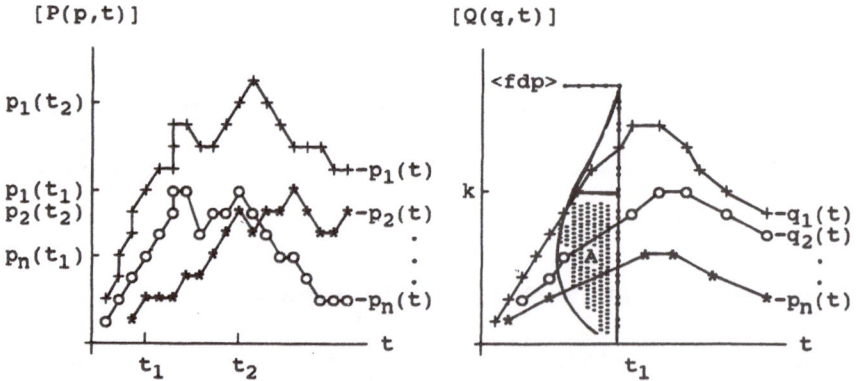

Fig. 3.53 Graphical representation of a stochastic process

$$A = \text{Prob}\{[Q(q, t_1)] \leqq k\}$$

$$[P(p, t)] = :[p_1(t), p_2(t), \ldots p_n{}^{(t)}]$$

$$[Q(q, t)] = :[q_1(t), q_2(t), \ldots q_n(t)].$$

The sample functions of the effective rain and of the surface runoff forming random processes (stochastic processes) are considered preliminarily as functions of continuous variables and later on of discrete variables.

The random process is considered as a function of two variables, one of which represents the time and is of the analytical type. The other is a random variable depending on a probability distribution [The dependence on a probability density function ⟨pdf⟩ is indicated by square brackets]. The stochastical behaviour of a random variable is described by a probability density function of the first order, and that of a random process is completely described by infinitely many joint probability density functions. The density function is characterized by statistical measures or parameters.

Some of the parameters often used in the stochastic domain will be explained as follows:

Mean value, (expected value):
$$E[P(p \cdot t)] = \int_{-\infty}^{+\infty} p \cdot \langle pdf \rangle (p, t) \, dp \qquad (3.271)$$

$$E[Q(q, t)] = \int_{-\infty}^{+\infty} q \cdot \langle pdf \rangle (q, t) \, dq \qquad (3.272)$$

Correlation:
$$Rpq(t_i, t_j) = \int_{-\infty}^{+\infty} \int_{-\infty}^{+\infty} pq \cdot \langle pdf \rangle (p, t_i; q, t_j) \, dp \, dq \qquad (3.273)$$

Autocorrelation:
$$Rpp(t_i, t_j) = \int_{-\infty}^{+\infty} \int_{-\infty}^{+\infty} pp \cdot \langle pdf \rangle (p, t_i; p, t_j) \, dp \, dp \qquad (3.274)$$

$$Rqq(t_i, t_j) = \int_{-\infty}^{+\infty} \int_{-\infty}^{+\infty} qq \cdot \langle pdf \rangle (q, t_i; q, t_j) \, dq \, dq. \qquad (3.275)$$

In order to determine these stochastical parameters, the probability density functions of the processes considered have to be known, and the ensemble of the random functions must also be known.

For practical calculations, these formulas are not useful because the measurements in nature will show only sample functions, whereas the totality of the process remains unknown. Moreover, the rainfall-runoff events can be recorded only in successive time sequences so that the sample functions are without recurrence.

Therefore, without essential restrictions the rainfall and the runoff process cannot be stochastically described.

Some of these restrictions are:

1. Requirement for stationarity

A stochastic process is called stationary if its statistical properties (parameters) are invariant with time. This means that the input process and the output process as well as the input–output relations do not change with time. In particular, a stochastic process will be called stationary if its mean is constant (wide-sense stationary).

2. Requirement for ergodicity

A stochastic process is called ergodic if its statistical parameters of the ensemble are equal to that of time. This means that the statistical parameters of a stochastic process determined from the ensemble at a certain time are equal to those determined from a single sample function observed over the whole time.

In detail, a stochastic process will be called ergodic if its ensemble averages are equal to the time averages. Obviously, an ergodic process must be stationary. However, a stationary process is not necessarily ergodic.

By these constraints it is possible to gather successive time measurements of rainfall or of discharge to an ensemble forming a random process.

Therefore, some of the parameters of ergodic stochastic processes can be determined as follows:

Mean:
$$E[P(p \cdot t)] = \int_{-\infty}^{+\infty} p \cdot \langle pdf \rangle (p, t) \, dp \quad \text{(expected value)} \tag{3.276}$$

$$= \lim_{T \to \infty} 1/2T \int_{-T}^{+T} p(t) \, dt = \text{const.} \tag{3.277}$$

Correlation:
$$Rpq(t_i, t_j) = \int_{-\infty}^{+\infty} \int_{-\infty}^{+\infty} pq \cdot \langle pdf \rangle (p, t_i; q, t_j) \, dp \, dq \tag{3.278}$$

$$= \lim_{T \to \infty} 1/2T \int_{-T}^{+T} p(t) q(t \pm \tau) \, dt, \tag{3.279}$$

$$\text{with } (t_j - t_i) = \tau \text{ und } Rpq(\tau) = Rqp(-\tau)$$

Autocorrelation:
$$Rpp(t_i, t_j) = \int_{-\infty}^{+\infty} \int_{-\infty}^{+\infty} pp \cdot \langle pdf \rangle (p, t_i; p, t_j) \, dp \, dp$$

$$= \lim_{T \to \infty} 1/2T \int_{-T}^{+T} p(t) p(t \pm \tau) \, dt, \tag{3.280}$$

$$\text{with } (t_j - t_i) = \tau \text{ und } Rpp(\tau) = Rpp(-\tau). \tag{3.281}$$

Another constraint often implied is the requirement of normality. This means that the considered process is normally distributed and its random variables have a Gaussian or normal probability density function (n-dimensional).

We shall now deal with a linear filter to model the relationship between the stochastic processes of precipitation and that of runoff in a catchment area.

$$[P(p, t)] \Rightarrow \boxed{h(t)} \Rightarrow [Q(q, t)].$$

With $[P(p, t)] = p(t)$ and $[Q(q, t)] = q(t)$, the linear system is represented by the convolution integral as follows:

$$q(t) = \int_0^t h(t)p(t - \tau)\,d\tau. \tag{3.282}$$

By substitution, this convolution integral for the random variables in the equations, for determination of the statistical parameters are obtained:

1. Mean value:

$$E[Q(q, t)] = E\left[\int_{-\infty}^{+\infty} h(\tau)p(t - \tau)\,d\tau\right] = \int_{-\infty}^{+\infty} h(\tau)E[p(t - \tau)]\,d\tau \tag{3.283}$$

2. Correlation:

a) Autocorrelation:

$$Rqq(t_i, t_j) = E[Q(q, t_i)Q(q, t_j)] \tag{3.284}$$

$$E\left[\int_{-\infty}^{+\infty} h(\tau)p(t_i - \tau)\,d\tau \int_{-\infty}^{+\infty} h(\mu)p(t_j - \mu)\,d\mu\right] \tag{3.285}$$

$$= \int_{-\infty}^{+\infty}\int_{-\infty}^{+\infty} h(\tau)h(\mu)E[p(t_i - \tau)p(t_j - \mu)]\,d\tau\,d\mu \tag{3.286}$$

$$= \int_{-\infty}^{+\infty}\int_{-\infty}^{+\infty} h(\tau)h(\mu)Rpp(t_i - \tau, t_j - \mu)\,d\tau\,d\mu \tag{3.287}$$

with $\tau - \mu = \sigma$ yields:

$$= \int_{-\infty}^{+\infty}\int_{-\infty}^{+\infty} h(\sigma + \mu)h(\mu)Rpp(t_i - \mu - \sigma, t_j - \mu)\,d\mu\,d\sigma. \tag{3.288}$$

In the electrical theory of circuits the function $g(\sigma)$ is known as the filter correlation function of Lampard (1955):

$$g(\sigma) = \int_{-\infty}^{+\infty} h(\mu)h(\sigma + \mu)\,d\mu. \tag{3.289}$$

Substitution of this relation in the formula before last yields:

$$Rqq(t_i, t_j) = \int_{-\infty}^{+\infty} g(\sigma)Rpp(t_i - \mu - \sigma, t_j - \mu)\,d\sigma. \tag{3.290}$$

For stationary linear systems can be written:
$(t_i - \mu) - (t_j - \mu) = u$, and we obtain:

$$Rqq(u) = \int_{-\infty}^{+\infty} g(\sigma)Rpp(u - \sigma)\,d\sigma \tag{3.291}$$

b) Correlation:

$$Rqp(t_i, t_j) = E[Q(q, t_i)P(p, t_j)] \tag{3.292}$$

$$= E\left[\int_{-\infty}^{+\infty} h(\tau)p(t_i - \tau)p(t_j)\,d\tau \right] \tag{3.293}$$

$$\boxed{Rqp(u) = \int_{-\infty}^{+\infty} h(\tau)\,Rpp(u - \tau)\,d\tau} \tag{3.294}$$

or

$$Rqq(t_i, t_j) = E[Q(q, t_i)Q(q, t_j)] \tag{3.295}$$

$$= E\left[\int_{-\infty}^{+\infty} h(\tau)p(t_i - \tau)q(t_j)\,d\tau \right] \tag{3.296}$$

$$\boxed{Rqq(u) = \int_{-\infty}^{+\infty} h(\tau)\,Rpq(u - \tau)\,d\tau} \tag{3.297}$$

3.7.2   Equations Wiener-Hopf and Yule-Walker

The linear system theory will be applied to the hydrological relation between the stochastic processes of precipitation $p(t)$ and that of runoff $q(t)$ in a catchment area.

By the convolution integral this relationship can be mathematically formulated as follows:

$$q(t) = \int_{-\infty}^{+\infty} h(\tau)p(t - \tau)\,d\tau. \tag{3.298}$$

In the case of the unknown system response function $h(t)$, the integral equation has to be resolved, where the stochastic in- and output processes have to be known. In this calculation, well known as an identification of the system, a deterministical function must be made out from two recorded stochastic processes.

Using the concept of linear mean squares estimation, it can be shown that in the convolution integral formula deterministical functions can be substituted for the two stochastic processes. They represent the statistical properties of both stochastic processes.

The mean squares error between the measured and calculated output process will be minimum.

$$EF = E\left[\left| q(t) - \int_{-\infty}^{+\infty} h(\tau)p(t - \tau)\,d\tau \right|^2 \right] \rightarrow \min. \tag{3.299}$$

To minimize the error, we differentiate the error function in relation to the unknown $h(t_i)$ and equal to zero:

$$d(EF)/d[h(t_i)] = E\left\{2\left[q(t) - \int_{-\infty}^{+\infty} h(\tau)p(t-\tau)d\tau\right][-p(t_i-\tau)]\right\} = 0$$

(3.300)

or

$$-E[q(t)p(t_i)] + \int_{-\infty}^{+\infty} h(\tau)E[p(t-\tau)p(t_i-\tau)]d\tau = 0.$$

(3.301)

With $t - t_i = u$ we obtain:

$$-Rpq(u) + \int_{-\infty}^{+\infty} h(\tau)Rpp(u-\tau)d\tau = 0$$

(3.302)

$$\boxed{Rqp(u) = \int_{-\infty}^{+\infty} h(\tau)Rpp(u-\tau)d\tau.}$$

(3.303)

This equation consists of only deterministical functions, where the stochastic input process p(t) is characterized by its autocorrelation and the stochastic output process q(t) by the cross-correlation function. The equation will be used to determine the system response function h(t).

This equation can also be obtained from the consideration of the parameters of linearily related stochastic processes, as has already been shown in Section 3.7.1.

However, this equation will be seen to be more important in other respects, as it is a necessary and sufficient condition to determine the system response function with a minimal mean squares error between the linear system model and the real measured runoff data.

This equation is known by the name of non-causal Wiener filter. (non-causal because the limits of the integral are from $-\infty$ to $+\infty$).

For causal systems the limits of the integral are from 0 to $+\infty$, and this equation is known by the name of the Wiener–Hopf integral equation.

Using the Wiener filter model, the arising error can be determined as follows:

$$Error = E\left[\left|q(t) - \int_{-\infty}^{+\infty} h(\tau)p(t-\tau)d\tau\right|^2\right] = Rqq(0) - \int_{-\infty}^{+\infty} h(\tau)Rpq(\tau)d\tau.$$

(3.304)

Using discretized variables the linear system relationship between the involved stochastic processes can be written in the following form of a sum:

$$\boxed{q(n) = \sum_{k=-N}^{k=+N} h(k)p(n-k)} \quad \text{for } N \to \infty$$

(3.305)

$$q(0) = h(-N)P(0+N) + \cdots + h(-1)p(1) + h(0)p(0) + h(+1)p(-1) + \cdots + h(N)p(0-N)$$
$$q(1) = h(-N)p(1+N) + \cdots + h(-1)p(2) + h(0)p(1) + h(+1)p(0) + \cdots + h(N)p(1-N)$$

$$q(n) = h(-N)p(n+N). + h(-1)p(n+1) + h(0)p(n) + h(+1)p(n-1) + \cdot + h(N)p(n-N).$$

For the mean value there is obtained:

$$E[q(n)] = E\left[\sum_{-\infty}^{+\infty} h(k)p(n-k) \right] = \sum_{-\infty}^{+\infty} h(k)E[p(n-k)]. \qquad (3.306)$$

For the correlation the result is:

$$Rqp(m) = \sum_{-\infty}^{+\infty} h(k)\,Rpp(m-k). \qquad (3.307)$$

For causal stationary systems yields:

$$Rqp(m) = \sum_{k=0}^{m} h(k)\,Rpp(m-k) \qquad (3.308)$$

$$Rqp(0) = h(0)Rpp(0) + h(1)Rpp(-1) + h(2)Rpp(-2) + \cdots + h(m)Rpp(0-m)$$
$$Rqp(1) = h(0)Rpp(1) + h(1)Rpp(0) + h(2)Rpp(-1) + \cdots + h(m)Rpp(1-m)$$
$$Rqp(2) = h(0)Rpp(2) + h(1)Rpp(1) + h(2)Rpp(0) + \cdots + h(m)Rpp(2-m)$$

$$Rqp(m) = h(0)Rpp(m) + h(1)Rpp(m-1) + h(2)Rpp(m-2) + \cdots + h(m)Rpp(m-m).$$

This system of linear equations to determine the unknown quantities h(k) is called the Yule-Walker equation. It is the analogue of the Wiener integral equation for discretized variables.

Physical Constraints and the Wiener–Hopf Integral Equation

The variety of functions satisfying the following equation:

$$Rqp(u) = \int_{0}^{+\infty} h(\tau)Rpp(u-\tau)\,d\tau \qquad (3.309)$$

is larger than the quantity of system response functions that can be realized in a physical system.

If $u < 0$, the cross-correlation and autocorrelation function show any values, and consequently, the system response function h(t) will also show any values for $t < 0$.

However, the condition for stable hydrological systems requires that the system response function has to be without negative values (see Sect. 3.6.2).

Using the Laplace transformation, the Wiener integral equation can be written by its transforms:

$$Sh(s) = Sqp(s)/[Spp(s)Spp(-s)], \quad \text{where } s = \text{complex number,}$$

with £[Rqp(u)] = Sqp(s)
        £[h(t)] = Sh(s)   and
        £[Rpp(σ)] = Spp(s).

For $t < 0$ only the retransform of Spp($-s$) will possibly show negative values different from zero that contribute to the system response. Therefore, the transformed integral equation has to be of such a form that at retransformation there is no contribution of negative values different from zero for $t < 0$.

After the splitting of the transform in the functions Spp(s) and Spp($-s$), and only retransforming that part containing the function Spp($-$), and a Laplace transformation again (but one-sided) results in an equation where the system response function shows no negative values for $t < 0$.

This procedure is presented by the following equations:

Splitting:

$$Sh(s) = 1/Spp(s) * Sqp(s)/Spp(-s) = 1/Spp(s) * £[h1(s)] \tag{3.310}$$

Retransformation:

$$h_1(t) = 1/2\pi i \int_{-\infty}^{+\infty} [Sqp(s)/Spp(-s)] \exp(st) \, ds. \tag{3.311}$$

One-sided Laplace transformation:

$$h_2(s) = \int_0^{+\infty} h_1(t) \exp(-st) \, dt. \tag{3.312}$$

Substitution:

$$Sh(s) = 1/Spp(s) \int_0^{+\infty} (1/2\pi i \int_{-\infty}^{+\infty} [Sqp(u)/Spp(-u)] \exp(ut) du) \exp(-st) \, dt. \tag{3.313}$$

The equation obtained can be analytically evaluated by means of the calculus of the residues. However, for particular practical problems a simpler method is used for a calculation demonstrated later.

3.7.3 Common Random Processes as System Input

Stochastic processes are characterized by the statistical parameters of the distribution function. The observation of phenomena in nature shows different types depending on development in time. Two of the commonly used stochastic processes in the theory of linear systems in hydroscience are the processes of white noise and of Markoff.

Process of White Noise

The random process of white noise [w(t)] is characterized by the following statistical properties:

1. The autocorrelation of the process is zero for all t_i, t_j with $t_i \neq t_j$.

   Stationary white noise: $\quad Rww(\tau) = 0 \quad$ with $\tau = t_i - t_j$
   $$Rww(0) = Ko = const.$$
   $$Rww(\tau) = Ko\delta(\tau)$$

   Non-stationary white noise: $Rww(\tau) = 0 \quad$ with $\tau = t_i - t_j$
   $$Rww(0) = K(t)$$
   $$Rww(\tau) = K(t)\delta(\tau)$$

2. The averages are constant: $E[w(t)] = K \quad$ with $K \in R$

The process is sketched for demonstration as follows:

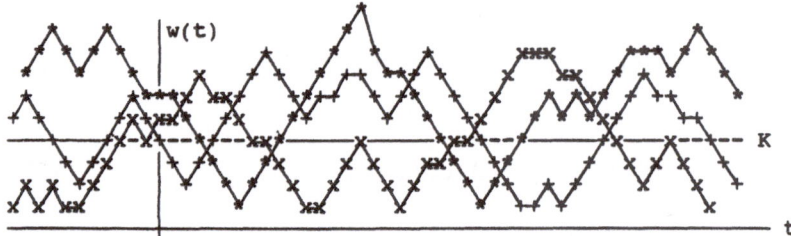

Fig. 3.54 Graphical representation of the white noise

Assuming that the precipitation process in a hydrological system can be represented by a stochastic process of white noise, the system response function can be determined by means of the Wiener filter model. Both the autocorrelation function of the runoff process as well as the cross-correlation between the precipitation process of white noise and the runoff process can be used to calculate the system response.

The non-causal Wiener filter model is represented by the following equation:

$$q(t) = \int\limits_{-\infty}^{+\infty} h(\tau)w(t-\tau)\,d\tau. \tag{3.314}$$

With $Rpp(t) = Rww(t) = K_0\delta(t)$ we obtain:

1.

$$Rqw(u) = \int\limits_{-\infty}^{+\infty} h(\tau)Rww(u-\tau)\,d\tau = K_0 \int\limits_{-\infty}^{+\infty} h(\tau)\delta(u-\tau)\,d\tau = K_0 h(u) \tag{3.315}$$

$$h(u) = Rqw(u)/K_0. \tag{3.316}$$

2.

$$Rqq(u) = \int\limits_{-\infty}^{+\infty} g(\sigma)Rww(u-\sigma)\,d\sigma = K_0 \int\limits_{-\infty}^{+\infty} g(\sigma)\delta(u-\sigma)\,d\sigma = K_0 g(u) \quad (3.317)$$

with

$$g(\sigma) = \int\limits_{-\infty}^{+\infty} h(\mu)h(\sigma+\mu)\,d\mu \quad \text{yields:} \tag{3.318}$$

$$Rqq(u)/K_0 = \int\limits_{-\infty}^{+\infty} h(\mu)h(u+\mu)\,d\mu. \tag{3.319}$$

Example for demonstrating the method of calculation:

Given   : Effective rain as a stationary random process of white noise with an autocorrelation of $Rpp(\tau) = Rww(\tau) = \delta(\tau)$.
Runoff as a stationary random process with an autocorrelation of $Rqq(\tau) = \exp(-a|\tau|)$, $a > 0$.
(Calculation is carried out by use of the computer program PCORR, Appendix 15).

Desired : System response function h(t)

Solution: The non-causal Wiener filter model is used.

$$Rqq(u) = \int\limits_{-\infty}^{+\infty} g(\sigma)Rww(u-\sigma)\,d\sigma = \int\limits_{-\infty}^{+\infty} g(\sigma)\delta(u-\sigma)\,d\sigma = g(u), \tag{3.320}$$

with

$$g(\sigma) = \int\limits_{-\infty}^{+\infty} h(\mu)h(\sigma+\mu)\,d\mu \tag{3.321}$$

yields:

$$\exp(-a|t|) = \int\limits_{-\infty}^{+\infty} h(\mu)h(t+\mu)\,d\mu. \tag{3.322}$$

From the Laplace transformation follows:

$$\int\limits_{-\infty}^{+\infty} \exp(-a|t|)\exp(-st)\,dt = \int\limits_{-\infty}^{+\infty}\left[\int\limits_{-\infty}^{+\infty} h(\mu)h(t+\mu)\,d\mu\right]\exp(-st)\,dt \tag{3.323}$$

$$\int\limits_{-\infty}^{0} \exp(at)\exp(-st)\,dt + \int\limits_{0}^{+\infty} \exp(-at)\exp(-st)\,dt$$

$$= \int\limits_{-\infty}^{+\infty} h(\mu)\,d\mu \int\limits_{-\infty}^{+\infty} h(t+\mu)\exp(-st)\,dt \tag{3.324}$$

$$\int\limits_{-\infty}^{0} \exp[(a-s)t]\,dt + \int\limits_{0}^{+\infty} \exp[-(a+s)t]\,dt$$

$$= \int\limits_{-\infty}^{+\infty} h(\mu)\exp(s\mu)\,d\mu \int\limits_{-\infty}^{+\infty} h(\sigma)\exp(-s\sigma)\,d\sigma \tag{3.325}$$

$$1/(a-s) + 1/(a+s) = \pounds[h(-\mu)] * \pounds[h(\sigma)] \tag{3.326}$$

$$2a/(a^2 - s^2) = \pounds[h(-\mu)] * \pounds[h(\sigma)] \tag{3.327}$$

$$[\sqrt{2a}/(a-s)][\sqrt{2a}/(a+s)] = \pounds[h(-\mu)] * \pounds[h(\sigma)] \tag{3.328}$$

$$[\sqrt{2a}/(a+s)] = \qquad \pounds[h(\sigma)] \tag{3.329}$$

$$\sqrt{2a}\exp(-at) = h(t). \tag{3.330}$$

Using the constraint $\int_0^\infty h(t)\,dt = 1$ we can determine the factor a, and obtain: $a = 2$.
For $Rqq(t) = \exp(-2|t|)$ yields the system response function:

$$h(t) = 4\exp(-2t). \tag{3.331}$$

Markoff Process

A Markoff process is characterized by a conditional probability density function that has dependencies only on values for the last time but not for earlier times. In other words, it is a process whose past has no influence on the future if its present is determined.

The characteristic of the Markoff process can be mathematically formulated as follows:

$$\text{Prob.}(x_n, t_n | x_1, t_1, x_2, t_2, \ldots, x_{n-1} t_{n-1}) = \text{Prob.}(x_n, t_n | x_{n-1} t_{n-1}). \tag{3.332}$$

Concerning the expected value E (average) it can be written:

$$E(x_n, t_n | x_1, t_1, x_2, t_2, \ldots, x_{n-1} t_{n-1}) = E(x_n, t_n | x_{n-1} t_{n-1}). \tag{3.333}$$

A stationary random process $[p(t)]$ with Gaussian probability density function can be a Markoff process only if the autocorrelation function is exponential:

$$Rpp(\tau) = Rpp(0)\exp(-a|\tau|) \tag{3.334}$$

with (a) as a parameter. This is a special property of stationary Markoff processes with a normal distribution. (For proof refer to Schwarz and Friedland 1965, pp. 299)

Assuming that the effective rain and runoff in a hydrological system are stationary random processes normally distributed, the autocorrelation function of both of the processes can be presented by exponential functions of different parameters. The system response function h(t) can be determined by means of the Wiener filter as follows:

$$Rqq(u) = \int_{-\infty}^{+\infty} g(\sigma)Rpp(u-\sigma)d\sigma \quad \text{with} \quad g(\sigma) = \int_{-\infty}^{+\infty} h(\mu)h(\sigma+\mu)d\mu. \tag{3.335}$$

Example to demonstrate the method of calculation:

Given : Effective rain as a stationary random process normally distributed and its autocorrelation:

$$Rpp(\tau) = \alpha \cdot \exp(-a|\tau|).$$

Runoff as a stationary Markoff process normally distributed with autocorrelation:

$$Rqq(\tau) = \beta \cdot \exp(-b|t|)$$
$$a, b, \alpha, \beta > 0.$$

Desired : System response function h(t)

Solution: The non-causal Wiener filter model is used.

$$Rqq(u) = \int_{-\infty}^{+\infty} g(\sigma)Rpp(u-\sigma)d\sigma \quad \text{with} \quad g(\sigma) = \int_{-\infty}^{+\infty} h(\mu)h(\sigma+\mu)d\mu \quad (3.336)$$

$$\beta \cdot \exp(-b|t|) = \int_{-\infty}^{+\infty} h(\mu)h(t+\mu)d\mu \int_{-\infty}^{+\infty} \alpha \cdot \exp(-a|t-\sigma|)d\sigma. \quad (3.337)$$

From the Laplace transformation follows:

$$\beta \int_{-\infty}^{+\infty} \exp(-b|t|)\exp(-st)\,dt$$

$$= \alpha \int_{-\infty}^{+\infty}\left[\int_{-\infty}^{+\infty} h(\mu)h(t+\mu)\,d\mu \int_{-\infty}^{+\infty} \exp(-a|t-\sigma|)d\sigma\right]\exp(-st)\,dt \quad (3.338)$$

$$\beta/(b-s) + \beta/(b+s) = \pounds[h(-\mu)]*\pounds[h(\sigma)]*[\alpha/(a+s)+\alpha/(a-s)] \quad (3.339)$$

$$2\beta b/(b^2-s^2) = \pounds[h(-\mu)]*\pounds[h(\sigma)]*[2\alpha a/(a^2-s^2)]$$

$$\beta b(a^2-s^2)/[\alpha a(b^2-s^2)] = \pounds[h(-\mu)]*\pounds[h(\sigma)] \quad (3.340)$$

$$\sqrt{\beta b/\alpha a}(a+s)/(b+s) = \pounds[h(\sigma)] \quad (3.341)$$

$$\sqrt{\beta b/\alpha a}[a/(b+s)+1-b/(b+s)] = \pounds[h(\sigma)] \quad (3.342)$$

$$\sqrt{\beta b/\alpha a}[(a-b)\exp(-bt)+\delta(t)] = h(t). \quad (3.343)$$

Using the constraint $\int_0^\infty h(t)\,dt = 1$ we can determine the coefficients by the equation: $\alpha \cdot b = \beta \cdot a$, and with: $a > b$, we obtain with $a = 0.8$, $b = 0.5$ and $\alpha = 1$: $\beta = 0.625$.

Finally there results the system response function:

$$h(t) = 0.1875\exp(-0.5t) + 0.625\,\delta(t). \quad (3.344)$$

3.7.4 Random Processes of the ARMA Type

The effective rain [p(t)] or the runoff in a hydrological system is assumed to be a random process consisting of two additional parts of specified random processes.

One of these random processes is characterized by a specified mean value function and an autocorrelation function different from zero. This process is denoted by a signal process [s(t)]. The other additional random process is characterized by a constant mean and an autocorrelation of zero. This process is of white noise [w(t)].

The process of the effective rain can be mathematically formulated:

$$p(t) = s(t) + w(t). \tag{3.345}$$

The runoff process can be analogically written:

$$q(t) = \bar{s}(t) + w(t). \tag{3.346}$$

Using these processes as input in a linear system, the convolution integral can be applied for simulation [calculation of $q(t)$ from known $p(t)$] and forcasting [calculation of $q(t)$ from known $p(t-1)$, $p(t-2)$, ...].

Likewise it is possible to use the model for simulation and the forecasting only of rain or only of runoff.

The formulation of the model is as follows:

$$q(t) = \int_{-\infty}^{+\infty} h_1(\tau)s(t-\tau)d\tau + \int_{-\infty}^{+\infty} h_2(\sigma)w(t-\sigma)d\sigma = \int_{-\infty}^{+\infty} h_3(\alpha)p(t-\alpha)d\alpha. \tag{3.347}$$

In the following we consider the process of effective rain; however, for the runoff process the result will be analogous.

The aim is to obtain an equivalent of the process of effective rain by the process of white noise and to obtain a suitable system response function, so that:

$$p(t) \Leftrightarrow w(t), \quad \text{where} \quad s(t) = 0.$$

By means of the convolution we obtain:

$$q(t) = \int_0^{+\infty} h_2(\tau)w(t-\tau)d\tau = \int_0^{+\infty} h_3(\alpha)p(t-\alpha)d\alpha. \tag{3.348}$$

Using discretized variables the equation changes into:

$$q(n) = \sum_{k=0}^{\infty} h_1(k)p(n-k) = \sum_{m=0}^{\infty} h_2(m)w(n-m). \tag{3.349}$$

The equation can be rewritten in the form of a sum:

$$h_1(0)p(n) = -\sum_{k=1}^{\infty} h_1(k)p(n-k) + \sum_{m=0}^{\infty} h_2(m)w(n-m). \tag{3.350}$$

Substitution of $-h_1(k)/h_1(0) = a(k)$ and of $h_2(m)/h_1(0) = b(m)$ yields:

$$p(n) = \sum_{k=1}^{\infty} a(k)p(n-k) + \sum_{m=0}^{\infty} b(m)w(n-m) \tag{3.351}$$

or

$$\boxed{p(n) = a_1 p(n-1) + a_2 p(n-2) + \cdots + b_0 w(n-o) + b_1 w(n-1) + \cdots}$$

$$\tag{3.352}$$

A stochastic process satisfying this equation is called the ARMA process (autoregressive moving average).

For $a(k) = 0$ with $k = 1 - \infty$ and $b(m) \neq 0$ yields:

$$p(n) = b_0 w(n - o) + b_1 w(n - 1) + \cdots \tag{3.353}$$

A stochastic process satisfying this equation is called the MA process (moving average)

For $a(k) \neq 0$ and $b(m) = 0$ mit $m = 1 - \infty$ yields:

$$p(n) = a_1 p(n - 1) + a_2 p(n - 2) + \cdots + b_0 w(n - o) \tag{3.354}$$

A stochastic process satisfying this equation is called the AR process (auto-regressive)

The effective rain can be modelled by the processes ARMA, MA, or AR, if the constants $a_1, \ldots, a_k$ and $b_0, \ldots, b_m$ are known. The coefficients a_k and/or b_m can be determined by means of the method of mean squares estimation. The resulting system of linear equation can be solved if the coefficients of correlation are known. (Yule-Walker equations, cf. Sect. 3.7.2)

Based on the method of linear mean squares estimation and on a known autocorrelation function, the stochastic process of effective rain can be forecast. Depending on the number of times in the past and with regard to the influence on the forecast, the model is of different order.

3.7.5 Whitening and Innovation Filters

We are dealing now with a specific linear system whose input is a stochastic process $p(t)$ of rainfall or runoff analogically. For the property of the system response function $\Phi(t)$ it is required that the output of the system is a process of white noise $w(t)$ with mean of zero and, moreover, using the white noise process for system input, the output will be before the input process, and the system response is reciprocal to the system response before $\Omega(t) = 1/\Phi(t)$.

$$p(t) \rightarrow \boxed{\Phi(t)} \rightarrow w(t) \rightarrow \boxed{\Omega(t)} \rightarrow p(t)$$

$$w(t) = \int_{-\infty}^{+\infty} \Phi(\tau) p(t - \tau)\, d\tau \quad p(t) = \int_{-\infty}^{+\infty} \Omega(\tau) w(t - \tau)\, d\tau. \tag{3.355}$$

A linear system which such a specific system response $\Phi(t)$ is called a whitening filter, and a linear system with a system response function of $\Omega(t)$ is called an innovation filter.

The correlation functions have to be satisfied:

$$Rwp(u) = \int_{-\infty}^{+\infty} \Phi(\tau) Rpp(u - \tau)\, d\tau \quad \text{and:} \tag{3.356}$$

$$Rpw(u) = \int\limits_{-\infty}^{+\infty} \Omega(\tau)Rww(u-\tau)\,d\tau = \int\limits_{-\infty}^{+\infty} \Omega(\tau)\delta(u-\tau)\,d\tau = \Omega(u). \qquad (3.357)$$

With $Rpw(u) = Rwp(-u)$ and $Rww(u) = Rww(-u)$ is obtained:

$$\Omega(u) = \int\limits_{-\infty}^{+\infty} \Phi(\tau)Rpp(u-\tau)\,d\tau. \qquad (3.358)$$

Using the Laplace transforms the result will be:

$$£[\Omega(u)] = £[\Phi(\sigma)] * £[Rpp(\tau)] \qquad (3.359)$$

$$£[Rpp(\tau)] = £[\Omega(u)]/£[\Phi(\sigma)]. \qquad (3.360)$$

With $\Omega(t) = 1/\Phi(t)$ we obtain:

$$£[Rpp(\tau)] = £[\Omega(\tau)] * £[\Omega(-\tau)]. \qquad (3.361)$$

If the autocorrelation function of a stochastic process is real and known, it is always possible to find a system response function for a whitening filter. Therefore, an innovation filter can be substituted for each real stochastic process.

$$w(t) \rightarrow \boxed{\quad \Omega(t) \quad} \rightarrow p(t).$$

Example to demonstrate this substitution:

Given: Effective rain as a stationary stochastic process with autocorrelation
         function: $Rpp(\tau) = \exp(-a|\tau|)$
Desired: System response function $\Omega(t)$ for an innovation filter
Solution:

$$£[Rpp(u)] = £[\exp(-a|u|)] = 2a/(a^2 - s^2) = £[\Omega(t)] * £[\Omega(-t)] \qquad (3.362)$$

$$2a/(a^2 - s^2) = \sqrt{2a}/(a+b) * \sqrt{2a}/(a-b) = £[\Omega(t)] * £[\Omega(-t)] \qquad (3.363)$$

$$\sqrt{2a}/(a+b) = £[\Omega(t)] \qquad (3.364)$$

$$\sqrt{2a}\,\exp(-at) = \Omega(t). \qquad (3.365)$$

3.7.6   Filters of Smoothing and Prediction

As shown in Section 3.7.4, the ARMA model can be derived from the additivity of stochastic processes.

The processes of precipitation and runoff are assumed to consist of two stochastic components, the signal process and the white noise.

Mathematically formulated, we can write for precipitation:

$$p(t) = s(t) + w(t), \text{ and for runoff: } q(t) = \bar{s}(t) + w(t).$$

The processes $s(t)$ and $\bar{s}(t)$ are stochastic processes from different signals, and their stochastic parameters are differently specified, whereas the process of white noise is of the same component for precipitation as for runoff. Substituting the process of white noise in both the equations we obtain:

$$p(t) = q(t) + s(t) - \bar{s}(t) \quad \text{and} \quad q(t) = p(t) + \bar{s}(t) - s(t). \tag{3.366}$$

With $s(t) - \bar{s}(t) = v(t)$ the stochastic process of precipitation becomes:

$$p(t) = q(t) + v(t). \tag{3.367}$$

The difference $v(t)$ between the two stochastic signal processes is not contained in any complete signal, and therefore, it can be considered a noise process, whereas the process $q(t)$ completely containing the signal is now denoted as a signal process.

In the following the linear filter to estimate the process $q(t)$ (signal) is used in terms of the known process $p(t)$, and the noise process $v(t)$ such that the mean squares value of the resulting error [measured values minus estimated values of $q(t)$] is a minimum.

If the process $q(t)$ is estimated for times within an interval, whereas the process $p(t)$ is completely known, the model is called a smoothing filter.

In hydrology the smoothing filter can be used for simulating runoff or precipitation.

If the process $q(t)$ is estimated for times $t + \alpha$ outside the interval, whereas the process $p(t)$ is completely known, the model is called a prediction filter.

In hydrology the prediction filter can be used for forecasting runoff or precipitation.

The model is called a pure prediction filter if the noise process is zero [$v(t) = 0$], and in the case of noise [$v(t) \neq 0$] it is called a filtering model and a prediction.

The models can be mathematically formulated:

$$q(t) = \int_a^b h(\tau)p(t - \tau)\,d\tau \tag{3.368}$$

and specified as follows:

Smoothing filter: $a < t < b$ and $v(t) \neq 0$
Prediction filter: $t < a, \quad t > b$ and $v(t) = 0$ pure
                    $t < a, \quad t > b$ and $v(t) \neq 0$ + filtering.

In the following examples are given for more detailed consideration.

Smoothing Filter

If the process $p(t)$ is known for all times, the linear smoothing filter for estimating the process $q(t)$ is formulated by the following convolution integral:

$$q(t) = \int_{-\infty}^{+\infty} h(\tau)p(t - \tau)\,d\tau \quad \text{with } p(t) = [q(t) + v(t)]. \tag{3.369}$$

With regard to the requirement of a minimal mean squares error, we obtain the non-causal Wiener filter equation:

$$Rqp(u) = \int\limits_{-\infty}^{+\infty} h(\tau)Rpp(u-\tau)\,d\tau. \tag{3.370}$$

By Laplace transformation the solution results in:

$$\pounds[Rqp(u)] = \pounds[h(\tau)] * \pounds[Rpp(\sigma)] \tag{3.371}$$

$$h(t) = \pounds^{\wedge}\{\pounds[Rqp(u)]/\pounds[Rpp(\sigma)]\}. \tag{3.372}$$

The model has been reduced to the non-causal Wiener filter.

$$p(t) \rightarrow \boxed{\ h(t)\ } \rightarrow q(t) \quad \text{with } p(t) = q(t) + v(t)$$

and

$$Rpp(u) = Rqq(u) + Rqv(u) + Rvq(u) + Rvv(u).$$

In a special case where the processes $q(t)$ and $v(t)$ are asssumed to be uncorrelated, we obtain with $Rqv(u) = Rvq(-u) = 0$ and $Rqp(u) = Rqq(u)$:

$$h(t) = \pounds^{\wedge}[\pounds[Rqq(u)]/\{\pounds[Rqq(u)] + \pounds[Rvv(u)]\}]. \tag{3.373}$$

The autocorrelation functions are pair functions, and the Laplace transforms are functions of the complex number s^2.
Usually where the processes $q(t)$ and $v(t)$ can be correlated in some way, the solution results from:

$$\pounds[h(t)] = \pounds[Rqq(u) + Rvq(u)]/\pounds[Rqq(u) + Rqv(u) + Rvq(u) + Rvv(u)].$$

For the not vanishing of the cross-correlation functions, the Laplace transforms are functions of the complex number s.
Example for demonstrating the smoothing filter:
Given: Assuming the effective rain $p(t)$ and the corresponding runoff $q(t)$ are stochastic processes of the ARMA type such that:

$$p(t) = q(t) + v(t),$$

and the runoff process can be considered a signal.
Autocorrelation of the runoff: $Rqq(u) = \exp(-a|u|)$
Autocorrelation of the noise: $Rvv = k^2\delta(u)$
Uncorrelated signal and noise: $Rqv(u) = -Rvq(u) = 0$.
Desired: System response function $h(t)$ for a smoothing filter to estimate $q(t)$ if the precipitation $p(t)$ is known. (runoff simulation from precipitation known over all the times)
Solution: The linear filter can be formulated:

$$q(t) = \int\limits_{-\infty}^{+\infty} h(\tau)p(t-\tau)\,d\tau = \int\limits_{-\infty}^{+\infty} h(\tau)[q(t-\tau) + v(t-\tau)]\,d\tau. \tag{3.374}$$

The non-causal Wiener filter equation requires:

$$Rqp(u) = \int_{-\infty}^{+\infty} h(\tau) Rpp(u - \tau) d\tau, \text{ where:} \qquad (3.375)$$

$$Rpp(u) = Rqq(u) + Rqv(u) + Rvq(u) + Rvv(u) = Rqq(u) + Rvv(u)$$
$$Rqp(u) = Rqq(u) + Rqv(u) \qquad\qquad = Rqq(u).$$

The Laplace transformation yields:

$$\math£[h(t)] = \math£[Rqq(u)]/\{\math£[Rqq(u)] + \math£[Rvv(u)]\} \quad \text{or} \qquad (3.376)$$

$$\text{with} \quad \math£[h(t)] = Sh(s), \math£[Rqq(u)] = Sqq(s), \qquad (3.377)$$

$$\text{and} \quad \math£[Rvv(u)] = Svv(s) \quad \text{with } s = x + iy \qquad (3.378)$$

$$Sh(s) = Sqq(s)/[Sqq(s) + Svv(s)]. \qquad (3.379)$$

The constraints of stability require:

$$Shr(t) = 1/[Sqq(+s) + Svv(+s)] \int_{0}^{+\infty} h_1(t) \exp(-st) dt \quad \text{with} \qquad (3.380)$$

$$h_1(t) = 1/2\pi i \int_{-\infty}^{+\infty} \{Sqq(s)/[Sqq(-s) + Svv(-s)]\} \exp(st) ds. \qquad (3.381)$$

With $Sqq(s) = 2a/(a^2 - s^2)$ and $Svv(s) = k^2$ we obtain:

$$Sh(s) = 2a/(a^2 - s^2):[2a/(a^2 - s^2) + k^2] \qquad (3.382)$$

$$= 2a/(a^2 - s^2):[k^2(c^2 - s^2)/(a^2 - s^2)], \qquad (3.383)$$

where $c^2 = (2a + k^2 a^2)/k^2$.
Separating the poles for $+s$ and $-s$ terms yields:

$$Sh(s) = 2a/(a + s)(a - s):(a + s)(a - s)/k(c + s)k(c - s).$$

Rewriting this term yields the following result:

$$Sh(s) = (a + s)/k(c + s) \cdot 2a/k(a + s)(c - s). \qquad (3.384)$$

1. Solution by the calculus of residues:

$$h_i(t) = 1/2\pi i \int_{-\infty}^{+\infty} [2a/k(a + s)(a - s)] \exp(st) ds. \qquad (3.385)$$

Forming the sum of residues:

$$h_1(t) = \lim_{s \to -a} (a + s)[2a/k(a + s)(c - s)] \exp(st)$$

$$+ \lim_{s \to c} (c - s)[2a/k(a + s)(c - s)] \exp(st) \qquad (3.386)$$

$$h_1(t) = [2a/k(c + a)] \exp(-at) + [2a/k(a + c)] \exp(ct). \qquad (3.387)$$

Only considering the stable part for solution:

$$h_{1r}(t) = [2a/k(c + a)] \exp(-at). \qquad (3.388)$$

The Laplace transformation of the stable part gives:

$$h_{2r}(s) = \int_0^{+\infty} [2a/k(c+a)] \exp(-at) \exp(-st) dt = [2a/k(c+a)]/(s+a).$$

$$(3.389)$$

Forming the complete transformed solution:

$$Sh_r(s) = (a+s)/k(c+s) \cdot [2a/k(c+a)]/(s+a) = 2a/k^2(c+a)(c+s). \quad (3.390)$$

Finally, we obtain the time function of the system response by retransformation:

$$h(t) = [2a/k^2(c+a)] \exp(-ct). \tag{3.391}$$

2. Solution by partial fractioning:

$$2a/(a+s)k(c-s) = A/(a+s) + B/k(c-s). \tag{3.392}$$

Comparing the coefficient yields:

$$Ak(c-s) + B(a+s) = Akc - Aks + Ba + Bs \tag{3.393}$$

$$Akc + Ba = 0$$

$$-Aks + Bs = 2a \Rightarrow A = 2a/k(c+a). \tag{3.394}$$

Considering only the stable part for solution we obtain:

$$Sh_r(s) = (a+s)/k(c+s) \cdot [2a/k(c+a)]/(s+a) \tag{3.395}$$

Pure Prediction

For estimation of the process $q(t+\alpha)$ at times α in terms of the known process $q(t)$ a linear filter is used formulated as follows:

$$q(t+\alpha) = \int_0^{+\infty} h(\tau)q(t-\tau) d\tau \quad \text{with } p(t) = q(t). \tag{3.396}$$

The determination of $h(t)$ by the method of least squares requires solving the Wiener–Hopf integral equation:

$$Rqq(u+\alpha) = \int_0^{+\infty} h(\tau)Rqq(u-\tau) d\tau \quad \text{with } u \geq 0. \tag{3.397}$$

Using the transform technique it is sometimes difficult to satisfy the constraints for stability, because the equation only holds for $u \geq 0$, but $Rqq(u)$ does not vanish for $u \leq 0$. However, the solution can be established using the calculus of residues (see example for smoothing).

In a simpler way the integral equation can be solved by means of an innovation filter replacing the process $q(t)$.

If the linear system is causal and stationary it is always possible to find an innovation filter for system replacing.

$$w(t) \rightarrow \boxed{\Omega(t)} \rightarrow \Leftrightarrow q(t).$$

The innovation filter can be mathematically formulated:

$$q(t) = \int\limits_0^{+\infty} \Omega(\tau)w(t-\tau)\,d\tau. \tag{3.398}$$

In this equation the system function $\Omega(t)$ and the white noise $w(t)$ are known. They are used as substitutes for the process $q(t)$.
The filter for prediction can be written as follows:.

$$q(t+\alpha) = \int\limits_0^{+\infty} \Omega(\tau)w(t+\alpha-\tau)\,d\tau. \tag{3.399}$$

With $\beta = \tau - \alpha$ is obtained:

$$q(t+\alpha) = \int\limits_0^{+\infty} \Omega(\beta+\alpha)w(t-\beta)\,d\beta. \tag{3.400}$$

In this prediction filter equation for the process $q(t+\alpha)$ the white noise is known, the system function $\Omega(\beta+\alpha)$ is certainly unknown, but the function $\Omega(\beta)$ is known by the calculation of the innovation filter. To obtain the function $\Omega(\beta+\alpha)$ the function $\Omega(\beta)$ has to be translated for the amount of α along the time axis.
The filter equation will be Laplace transformed:

$$\pounds[\Omega(t+\alpha)] = \int\limits_0^{+\infty} \Omega(t+\alpha)\exp(-st)\,dt. \tag{3.401}$$

With $\alpha + t = \tau$ then:

$$\pounds[\Omega(t+\alpha)] = \int\limits_0^{+\infty} \Omega(\tau)\exp[-s(\tau-\alpha)]\,d\tau \tag{3.402}$$

$$\pounds[\Omega(t+\alpha)] = \int\limits_0^{+\infty} \Omega(\tau)\exp(-s\tau)\exp(s\alpha)\,d\tau = \pounds[\Omega(t)]\cdot\exp(s\alpha). \tag{3.403}$$

The filter equation can be written in its Laplace transforms:

$$\pounds[q(t+\alpha)] = \pounds[\Omega(\alpha+t)]*\pounds[w(t)] = \pounds[\Omega(t)]\cdot\exp(s\alpha)*\pounds[w(t)]. \tag{3.404}$$

The whitening filter can also be written in its Laplace transforms:

$$\pounds[w(t)] = \pounds[\Phi(\sigma)]*\pounds[q(t)]. \tag{3.405}$$

By substitution follows:

$$\pounds[q(t+\alpha)] = \pounds[\Omega(t)]\exp(s\alpha)*\pounds[\Phi(\sigma)]*\pounds[q(t)] = \pounds[h(t)]*\pounds[q(t)]. \tag{3.406}$$

Finally, for the prediction filter the system function $h(t)$ can be determined by retransformation:

$$h(t) = \pounds^{\wedge}\{\pounds[\Omega(t)]\cdot\exp(s\alpha)*\pounds[\Phi(t)]\} \tag{3.407}$$

With $\Omega(t) = 1/\Phi(t)$ and $\pounds[Rqq(\tau)] = \pounds[\Omega(\tau)] * \pounds[\Omega(-\tau)]$

$$q(t) \rightarrow \boxed{\Phi(t)} \rightarrow \boxed{\Omega(t)\cdot\exp(s\alpha)} \rightarrow q(t+\alpha).$$

Example for demonstrating the method of calculation by use of continuous variables:

Given: The precipitation and runoff processes are assumed to be random processes of the ARMA type:

$$p(t) = q(t) + v(t) \tag{3.408}$$

Autocorrelationfunction of the
runoff process $q(t)$: $Rqq(u) = \exp(-a|u|)$
The noise process is zero: $v(t) = 0$

Desired: System function $h(t)$ for a system to predict the runoff $q(t+T)$

Solution: The innovation filter technique will be applied.

$$Rpp(u) = Rqq(u) + Rqv(u) + Rvq(u) + Rvv(u) = Rqq(u) \tag{3.409}$$

$$Rpq(u) = Rqq(u) + Rvq(u) \qquad\qquad = Rqq(u). \tag{3.410}$$

1. Calculation of the Laplace transform of $Spp(s) = \pounds[Rpp(u)]$,

$$Spp(s) = Sqq(s) = 2a/(a^2 - s^2) \quad \text{with } s = x + iy \tag{3.411}$$

$$Spq(s) = Sqq(s) = 2a/(a^2 - s^2). \tag{3.412}$$

2. Calculation of the innovation filter for replacing the process $p(t)$:

$$\pounds[\Phi(t)] = \sqrt{2a}/(a+s). \tag{3.413}$$

3. Calculation of the function:

$$hv(s) = \pounds[Rpq(u)]\exp(s\alpha)/\pounds[\Omega(-t)]: \tag{3.414}$$

$$hv(s) = 2a\cdot\exp(s\alpha)(a-s)/(a^2 - s^2)\sqrt{2a} \tag{3.415}$$

$$= 2a\cdot\exp(s\alpha)/(a+s)\sqrt{2a}. \tag{3.416}$$

4. Separating the part of the solution satisfying the constraints of stability by the partial fractioning:
(This point can be cancelled here because there is only one stable part of solution).

5. Determination of the result by multiplication the stable part with the inverse of the Laplace transform of the whitening filter system function

$$\pounds[h(t)] = \pounds[\Omega(t)] * hvr(s) \tag{3.417}$$

$$= \exp(s\alpha)2a(a+s)/2a(a+s) \tag{3.418}$$

$$h(t) = \exp(-a\alpha)\delta(t). \tag{3.419}$$

The prediction model can be formulated with the determined system function as follows:

$$q(t + \alpha) = \int_0^{+\infty} h(\tau)q(t - \tau)d\tau \tag{3.420}$$

$$= \int_0^{+\infty} \exp(-a\alpha)\delta(\tau)q(t - \tau)d\tau = \exp(-a\alpha)q(t). \tag{3.421}$$

Filtering and Prediction

We consider again the linear filter model to estimate the runoff process $q(t + \alpha)$ in terms of the precipitation process $q(t)$:

$$q(t + \alpha) = \int_0^{+\infty} h(\tau)p(t - \tau)d\tau \quad \text{with } p(t) = q(t) + v(t). \tag{3.422}$$

To determine the system function $h(t)$ by least squares estimation, the Wiener–Hopf integral equation has to be solved:

$$Rqp(u + \alpha) = \int_0^{+\infty} h(\tau)Rpp(u - \tau)d\tau \quad \text{with } u \geq 0. \tag{3.423}$$

If $p(t) = q(t)$ the system function can be determined for a prediction filter. For $\alpha = 0$ with $p(t) = q(t) + v(t)$ the system function can be determined for a smoothing filter.

As shown before, the solution of the Wiener–Hopf integral equation can be made in such a way that its limits of the integral will be extended from $-\infty$ to $+\infty$, then the equation can be solved by the transform technique, and consequently, only the stable part of the solution is considered. This is the so-called classic method.

Alternatively, for solution of the Wiener–Hopf equation an innovation filter can be used to replace the starting process. The determined system function of this replacing filter will be shifted about the prediction time interval.

In the following we shall be concerned with the innovation filter technique, and an example will be given for both demonstrating the method and comparing the result obtained by the classic manner of solution.

For real processes we can always find an innovation filter system replacing the precipitation process $p(t)$:

$$w(t) = \int_0^{+\infty} \Phi(\tau)p(t - \tau)d\tau \quad \text{or} \quad \pounds[w(t)] = \pounds[\Phi(\tau)] * \pounds[p(t)]. \tag{3.424}$$

The innovation filter $\Omega(t)$ for replacing the process $p(t)$ is formulated:

$$p(t) = \int_0^{+\infty} \Omega(\tau)w(t - \tau)d\tau \quad \text{or} \quad \pounds[p(t)] = \pounds[\Omega(\tau)] * \pounds[w(t)]. \tag{3.425}$$

By means of the identity where the transform of the known function $\Omega(t)$ is split off:

$$\pounds[Rpp(u)] = \pounds[\Omega(+t)] * \pounds[\Omega(-t)], \tag{3.426}$$

introduced in Section 3.7.5, the Wiener–Hopf equation can be written in the following form:

$$Rqp(u + \alpha) = \int_0^{+\infty} h(\tau)\, Rpp(u - \tau)\, d\tau = \int_0^{+\infty} h(\tau)[\Omega(+ u - \tau)\Omega(- u - \tau)]\, d\tau$$

$$(3.427)$$

with u > 0, and the transformed equation is:

$$£[Rqp(u + \alpha)] = £[h(t)] * £[\Omega(+ u)] * £[\Omega(- u)] \qquad \text{or} \qquad (3.428)$$

$$£[h(t)] = £[Rqp(u + \alpha)]/\{£[\Omega(+ u)] * £[\Omega(- u)]\} \qquad \text{or} \qquad (3.429)$$

$$£[h(t)] = 1/£[\Omega(+ u)] * £[Rqp(u + \alpha)]/£[\Omega(- u)] \qquad \text{or} \qquad (3.430)$$

$$£[h(t)] = £[\Phi(u)] * £[Rqp(u)]\exp(s\alpha)/£[\Omega(- u)] \qquad \text{or} \qquad (3.431)$$

$$h(t, \alpha) = £^{\wedge}[\{£[\Phi(- t)] * £[Rpq(u)]\} \cdot \exp(s\alpha) * £[\Phi(t)]] \qquad (3.432)$$

$$p(t) \rightarrow \boxed{\Phi(t)} \rightarrow \boxed{[\Phi(- t) * Rpq(u)] \cdot \exp(s\alpha)} \rightarrow s(t + \alpha).$$

Using the Z-transformation instead of the Laplace transformation, we analogically obtain:

$$h(n, \alpha) = Z^{\wedge}[\{Z[\Phi(- n)] * Z[Rsp(u)]\} \cdot z^{\alpha} * Z[\Phi(n)]]. \qquad (3.433)$$

This chapter will close with two examples demonstrating the method of calculation. In the first example continuous variables are used and discretized in the second.

1. Example for continuous variables:

Given   : Assuming the effective rain p(t) and the corresponding runoff q(t) are stochastical processes of the ARMA type such that:

$$p(t) = q(t) + v(t) \qquad (3.434)$$

and the runoff process can be considered a signal.

Autocorrelation of the runoff : $Rqq(u) = \exp(- a|u|)$
Autocorrelation of the noise   : $Rvv = k^2 \delta(u)$
Uncorrelated signal and noise: $Rqv(u) = - Rvq(u) = 0$

Desired : System response function h(t) for a smoothing and prediction filter to estimate q(t) if the precipitation p(t) is known

Solution: a) using the innovation filter technique:
         We have:

$$Rpp(u) = Rqq(u) + Rqv(u) + Rvq(u) + Rvv(u) = Rqq(u) + Rvv(u) \qquad (3.435)$$

$$Rpq(u) = Rqq(u) + Rvq(u) \qquad\qquad\qquad\quad = Rqq(u) \qquad (3.436)$$

1. Calculation of the Laplace transform of $Spp(s) = \mathcal{L}[Rpp(u)]$,

$$Spq(s) = \mathcal{L}[Rpq(u)]: \tag{3.437}$$

$$Spp(s) = Sqq(s) + Svv(s) = 2a/(a^2 - s^2) + k^2 \tag{3.438}$$

$$= k^2(c^2 - s^2)/(a^2 - s^2) \tag{3.439}$$

with $c^2 = (2a + k^2 a^2)/k^2$ where $s = x + iy$

$$Spq(s) = 2a/(a^2 - s^2). \tag{3.440}$$

2. Calculation of the innovation filter for replacing the process $p(t)$:

$$\mathcal{L}[\Phi(t)] = k(c + s)/(a + s). \tag{3.441}$$

3. Calculating the function:

$$hv(s) = \mathcal{L}[Rpq(u)]\exp(s\alpha)/\mathcal{L}[\Omega(-t)]: \tag{3.442}$$

$$hv(s) = 2a \cdot \exp(s\alpha)(a - s)/(a^2 - s^2)k(c - s) \tag{3.443}$$

$$= 2a \cdot \exp(s\alpha)/k(a + s)(c - s). \tag{3.444}$$

4. Separating the part of the solution satisfying the constraints for stability by partial fractioning:

$$\exp(s\alpha)[2a/k(a + s)(c - s)] = \exp(s\alpha)[A/(a + s) + B/k(c - s)]$$
$$\Rightarrow A = 2a/k(c + a) \tag{3.445}$$

$$hv_r(s) = \exp(s\alpha)2a/k(c + a)(a + s). \tag{3.446}$$

5. Determination of the result by multiplying the stable part with the inverse of the Laplace transform of the whitening filter system function.

$$\mathcal{L}[h(t)] = \mathcal{L}[\Omega(t)] * hv_r(s) \tag{3.447}$$

$$= \exp(s\alpha)2a(a + s)/k(c + s)k(c + a)(a + s) \tag{3.448}$$

$$= \exp(s\alpha)2a/k^2(c + a)(c + s) \tag{3.449}$$

$$h(t) = [2a/k^2(c + a)]\exp[-c(t + \alpha)]. \tag{3.450}$$

b) Classical method of solution:

$$q(t + \alpha) = \int_0^{+\infty} h(\tau)p(t - \tau)d\tau = \int_0^{+\infty} h(\tau)[q(t - \tau) + v(t - \tau)]d\tau. \tag{3.451}$$

For linear mean squares estimation of the system function $h(t)$ the Wiener–Hopf integral equation has to be satisfied:

$$Rqp(u + \alpha) = \int_0^{+\infty} h(\tau)Rpp(u - \tau)d\tau \quad \text{with } u \geq 0 \tag{3.452}$$

and:

$$Rpp(u) = Rqq(u) + Rqv(u) + Rvq(u) + Rvv(u) = Rqq(u) + Rvv(u) \qquad (3.453)$$

$$Rqp(u) = Rqq(u) + Rqv(u) \qquad\qquad\qquad = Rqq(u). \qquad (3.454)$$

The Laplace transforms are:

$$£[h(t)] = £[Rqq(u + \alpha)]/\{£[Rqq(u)] + £[Rvv(u)]\} \quad \text{or} \qquad (3.455)$$

$$£[h(t)] = Sh(s), \quad £[Rqq(u)] = Sqq(s), \quad \text{and} \quad £[Rvv(u)] = Svv(s) \qquad (3.456)$$

$$Sh(s) = Sqq(s)\exp(s\alpha)/[Sqq(s) + Svv(s)] \quad \text{with } s = x + iy. \qquad (3.457)$$

The condition for stability requires:

$$Sh_r(t) = 1/[Sqq(+s) + Svv(+s)] \int_0^{+\infty} h_1(t)\exp(-st)\,dt \quad \text{with} \qquad (3.458)$$

$$h_1(t) \doteq 1/2\pi i \int_0^{+\infty} \{Sqq(s)\exp(s\alpha)/[Sqq(-s) + Svv(-s)]\}\exp(st)\,ds \qquad (3.459)$$

For $Sqq(s) = 2a/(a^2 - s^2)$ and $Svv(s) = k^2$ we obtain:

$$Sh(s) = 2a\exp(s\alpha)/(a^2 - s^2):(2a/(a^2 - s^2) + k^2) \qquad (3.460)$$

$$= 2a\exp(s\alpha)/(a^2 - s^2):(k^2(c^2 - s^2)/(a^2 - s^2)) \qquad (3.461)$$

with $\quad c^2 = (2a + k^2 a^2)/k^2$.

Separating the poles in the complex half plane $(+s, -s)$ yields:

$$Sh(s) = 2a\exp(s\alpha)/(a + s)(a - s):(a + s)(a - s)/k(c + s)k(c - s). \qquad (3.462)$$

Rewriting this becomes:

$$Sh(s) = (a + s)/k(c + s) \cdot 2a\exp(a\alpha)/k(a + s)(c - s). \qquad (3.463)$$

The integral:

$$h_1(t) = 1/2\pi i \int_{-\infty}^{+\infty} [2a\exp(s\alpha)/k(a + s)(a - s)]\exp(st)\,ds \qquad (3.464)$$

is evaluated by the calculus of residues:

$$\dot{h}_1(t) = \lim_{s \to -a} (a + s)[2a\exp(s\alpha)/k(a + s)(c - s)]\exp(st)$$

$$+ \lim_{s \to c}(c - s)[2a\exp(s\alpha)/k(a + s)(c - s)]\exp(st) \qquad (3.465)$$

$$h_1(t) = [2a/k(c + a)]\exp[-a(t + \alpha)] + [2a/k(a + c)]\exp[c(t + \alpha)]. \qquad (3.466)$$

Only the stable part of the solution will be taken into account:

$$h_{1r}(t) = [2a/k(c + a)]\exp[-a(t + \alpha)]. \qquad (3.467)$$

Transforming this part by Laplace yields:

$$h_{2r}(s) = \int_0^{+\infty} [2a/k(c+a)]\exp[-a(t+\alpha)]\exp(-st)\,dt \tag{3.468}$$

$$= \exp(-a\alpha)[2a/k(c+a)]/(s+a). \tag{3.469}$$

Forming the complete solution, then:

$$Sh_r(s) = (a+s)/k(c+s)\cdot\exp(-a\alpha)[2a/k(c+a)]/(s+a) \tag{3.470}$$

$$= \exp(-a\alpha)2a/k^2(c+a)(c+s). \tag{3.471}$$

Finally, by retransformation we obtain:

$$h(t) = [2a/k^2(c+a)]\exp[-c(t+\alpha)]. \tag{3.472}$$

2. Example for discretized variables:

Given : Assuming the effective rain $p(n)$ and the corresponding runoff $q(n)$ are stochastic processes of the ARMA type such that:

$$p(n) = q(n) + v(n) \tag{3.473}$$

and the runoff process can be considered a signal.

Autocorrelation of the runoff : $Rqq(n) = 0.8^n$
Autocorrelation of the noise : $Rvv(n) = \delta(n)$
Uncorrelated signal and noise: $Rqv(u) = -Rvq(u) = 0$.

Desired : System response function $h(t)$ for a system able to estimate with minimal mean squares error:
a) the runoff process $q(n)$ in terms of the precipitation process $p(n)$ (smoothing), and
b) the runoff process $q(n+T)$ in terms of the precipitation process $p(n)$ (filtering and prediction)

Solution: Calculation of the Z-transforms of the correlation functions:

$$Z[Rqq(n)] = \sum_{n=-\infty}^{n=+\infty} 0.8^n z^{-n} = \sum_{n=-\infty}^{n=-\infty} 0.8^{-n} z^{-n} + \sum_{n=0}^{n=+\infty} 0.8^n z^{-n} \tag{3.474}$$

$$= \sum_{n=1}^{n=\infty} (0.8\,z)^n + \sum_{n=0}^{n=\infty} (0.8\,z^{-1})^n \tag{3.475}$$

$$= \sum_{n=0}^{n=\infty} (0.8\,z)^n + \sum_{n=0}^{n=\infty} (0.8\,z^{-1})^n - 1 \tag{3.476}$$

$$= 1/(1-0.8\,z) + 1/(1-0.8\,z^{-1}) - 1 \tag{3.477}$$

$$= 0.8z/(1-0.8z) + 1/(1-0.8z^{-1}) \tag{3.478}$$

$$= 0.36/(1-0.8z)(1-0.8z^{-1}) \tag{3.479}$$

$$Z[Rpp(n)] = Z[Rqq(n)] + Z[Rvv(n)] \tag{3.480}$$

$$= 0.36/(1 - 0.8z)(1 - 0.8z^{-1}) + 1 \tag{3.481}$$

$$= [0.36 + (1 - 0.8z)(1 - 0.8z^{-1})]/(1 - 0.8z)(1 - 0.8z^{-1}) \tag{3.482}$$

$$= (0.36 + 1 - 0.8z^{-1} - 0.8z + 0.64)/(1 - 0.8z)(1 - 0.8z^{-1}) \tag{3.483}$$

$$= (-0.8z^2 + 2z - 0.8)/z(1 - 0.8z)(1 - 0.8z^{-1}) \tag{3.484}$$

$$= [-0.8(z - 2)(z - 5)]/z(1 - 0.8z)(1 - 0.8z^{-1}) \tag{3.485}$$

$$= [-0.8(z - 2)(1 - 0.5z^{-1})]/(1 - 0.8z)(1 - 0.8z^{-1}) \tag{3.486}$$

$$= [0.8 \cdot 2(1 - 0.5z)(1 - 0.5z^{-1})]/(1 - 0.8z)(1 - 0.8z^{-1}) \tag{3.487}$$

$$= [1.6(1 - 0.5z)(1 - 0.5z^{-1})]/(1 - 0.8z)(1 - 0.8z^{-1}).$$

Calculation of the Z-transforms of the innovation filter:

$$Z[Rpp(u)] = \{Z[\Omega(+u)]\}\{Z[\Omega(-u)]\} \tag{3.488}$$

$$Z[\Omega(+u)] = \sqrt{1.6}(1 - 0.5z^{-1})/(1 - 0.8z^{-1}) = \sqrt{1.6}(z - 0.5)/(z - 0.8). \tag{3.489}$$

a) Smoothing Filter Model

$$Z[h(n)] = Z[Rqq(u)]/Z[Rpp(u)] \tag{3.490}$$

$$= Z[Rqq(u)]/\{Z[Rqq(u)] + Z[Rvv(u)]\} \tag{3.491}$$

$$= [0.36/(1 - 0.8z)(1 - 0.8z^{-1})]: \tag{3.492}$$

$$[1.6(1 - 0.5z)(1 - 0.5z^{-1})/(1 - 0.8z)(1 - 0.8z^{-1})].$$

Separating the stable part of the solution:

$$= (1 - 0.8z^{-1})/(1 - 0.5z^{-1}) \cdot 0.36/1.6(1 - 0.8z^{-1})(1 - 0.5z). \tag{3.493}$$

Partial fractioning of the stable part of the solution:

$$0.36/1.6(1 - 0.5z)(1 - 0.8z^{-1}) = Az/(1 - 0.5z) + B/(1 - 0.8z^{-1}) = \tag{3.494}$$

$$(Az - 0.8A + B - 0.5Bz)/(1 - 0.5z)(1 - 0.8z^{-1}). \tag{3.495}$$

Determination of the coefficients of A and B:

$$(A - 0.5B)z = 0 \qquad \Rightarrow A = 0.5B$$

$$B - 0.8A = 0.36/1.6$$

$$\Rightarrow B = 0.36/(1.6(1 - 0.4)) = 0.375; \Rightarrow A = 0.1875.$$

The complete solution yields:

$$Z[h(n)] = (1 - 0.8z^{-1})/(1 - 0.5z^{-1}) \cdot 0.375/(1 - 0.8z^{-1}) \tag{3.496}$$

$$= 0.375/(1 - 0.5z^{-1}) \tag{3.497}$$

$$h(n) = 0.375 \cdot 0.5^n. \tag{3.498}$$

With this solution the smoothing filter model can be written as follows:

$$q(n) = 0.375 \sum_{0}^{k = +\infty} 0.5^k p(n - k) \quad \text{or as AR(1) process:} \tag{3.499}$$

in the form of an autoregressive model (AR(1) process):

$$q(n) = 0.375 p(n) + 0.1875 p(n - 1). \tag{3.500}$$

b) Filtering and Prediction Model

The filter model and its Z-transforms are summarised in the following formula:

$$Z[h(n, \alpha)] = Z[\Phi(n)] * \{Z[\Phi(-n)] * Z[Rpq(u)]\} \cdot z^\alpha, \tag{3.501}$$

where is written:

$$Z[Rpq(u)] = Z[Rpp(u)] - Z[Rvv(u)] = Z[\Omega(+n)]Z[\Omega(-n)] - Z[Rvv(u)] \tag{3.502}$$

$$Z[\Phi(-n)] * Z[Rpq(u)] = Z[\Phi(-n)] * \{Z[\Omega(n)] * Z[\Omega(-n)] - Z[Rvv(u)]\} \tag{3.503}$$

$$= Z[\Omega(n)] - Z[\Phi(-n)] * Z[Rvv(u)]. \tag{3.504}$$

In the case: $\alpha = 0$ (smoothing): $Z[\Phi(-n)] = \Phi(0)$

$$Z[h(n)] = Z[\Phi(n)] * \{Z[\Omega(n)] - \Phi(0) * Z[Rvv(u)]\} \tag{3.505}$$

$$= 1 - \Phi(0) * Z[Rvv(u)] * Z[\Phi(n)] \tag{3.506}$$

$$= 1 - Z[Rvv(u)]/\{\Omega(0) * Z[\Omega(n)]\} \tag{3.507}$$

With: $Z[Rvv(u)] = 1$, $Z[\Omega(n)] = \sqrt{1.6}(z - 0.5)/(z - 0.8)$ and $\Omega(0) = \lim_{z \to \infty} Z[\Omega(n)]$, we obtain:

$$Z[h(n)] = 1 - (z - 0.8)/(\sqrt{1.6}) \cdot \sqrt{1.6}(z - 0.5) \tag{3.508}$$

$$= 1 - (z - 0.8)/[1.6(z - 0.5)] \tag{3.509}$$

$$= [1.6(z - 0.5) - (z - 0.8)]/1.6(z - 0.5) = 0.375z/(z - 0.5) \tag{3.510}$$

$$= 0.375/(1 - 0.5z^{-1}) \tag{3.511}$$

$$h(n) = 0.375 \, (0.5)^n. \tag{3.512}$$

As expected, the result is in accordance with that one of the smoothing filter model (because $\alpha = 0$).

In the case: $\alpha = 1$ (filtering and prediction), it is:

$$Z[\Phi(-n)] = \Omega(0) \tag{3.513}$$

$$Z[h(n)] = z - zZ[Rvv(u)]\Omega(0)Z[\Phi(n)] \tag{3.514}$$

$$= z - z\Omega(0)/Z[\Omega(0)] \tag{3.515}$$

$$= z - z\sqrt{1.6}(z - 0.8)/\sqrt{1.6}(z - 0.5) \tag{3.516}$$

$$= z - z(z - 0.8)/(z - 0.5) \tag{3.517}$$

$$= [z(z - 0.5) - z(z - 0.8)]/(z - 0.5) = 0.3z/(z - 0.5) \tag{3.518}$$

$$= 0.3z/(z - 0.5) = 0.3/(1 - 0.5z^{-1}) \tag{3.519}$$

$$h(t) = 0.3(0.5)^n. \tag{3.520}$$

With this solution the prediction filter model can be written in the following form:

$$q(n + 1) = 0.3 \sum_{k=0}^{k=\infty} 0.5^k p(n - k) \tag{3.521}$$

or in the form of an autoregressive model [AR(1) process]:

$$q(n + 1) = 0.3p(n) + 0.15p(n - 1). \tag{3.522}$$

4 Treatment of Hydrological Variables in Non-Linear Systems

In the background of the need for simulation or forecast modelling the catchment is considered as an engineering system that transforms rainfall input into streamflow output. In the mathematical methods shown above, this kind of application is possible only for linear systems. To apply linear models to problems in practice, the data have to be modified so that a linear behaviour can be approximated. Base-flow substraction or determination of the effective rain are used to obtain corresponding functions of a linear system. Leaving out of account the problem of obtaining accurate measurements of the involved quantities, the linear model does not represent the catchment behaviour, because the catchment behaves non-linearity. The first systematic treatment of the non-linear effects was reported by Minshall (1986), who demonstrated the dependency of the unit hydrograph on rain fall intensity. In accordance with the hydrodynamic equations describing hydrological processes, a non-linear behaviour of the catchment can be taken for granted. A useful non-linear representation of the rainfall runoff relationship is the second-order Volterra series model.

4.1 Volterra Series Model

A non-linear system can be described by the multinominal form of the convolution integral called the Volterra series. This series can be mathematically formulated:

$$Q(t) = \int_0^t h_1(\tau)p(t - \tau)\,d\tau + \int_0^t \int_0^t h_2(\tau_1, \tau_2)p(t - \tau_1)p(t - \tau_2)\,d\tau_1\,d\tau_2 \tag{4.1}$$

$$+ \cdots + \int_0^t \cdots \int_0^t h_n(\tau_1, \ldots, \tau_n)p(t - \tau_1) \cdots p(t - \tau_n)\,d\tau_1 \cdots d\tau_n. \tag{4.2}$$

The usefulness of the Volterra series for modelling surface runoff systems has been discussed by several scientists; for further studies refer to Amorocho and Orlob (1961); Diskin et al. (1972); Boneh (1971); Jacoby (1966).

Because of the fairly good results obtained by the use of linear models, the influence of non-linearity on modelling is assumed to be relatively small. Therefore, we consider it sufficient to add only a second term to the linear model, and we obtain a Volterra series model of the second order. Supposing stationarity, this model can be written in the following forms:

$$Q(t) = \int_0^t h_1(\tau)p(t-\tau)d\tau + \int_0^t\int_0^t h_2(\tau_1,\tau_2)p(t-\tau_1)p(t-\tau_2)\,d\tau_1\,d\tau_2 \qquad (4.3)$$

or

$$Q(t) = \int_0^t h_1(t-\tau)p(\tau)d\tau + \int_0^t\int_0^t h_2(t-\tau_1,t-\tau_2)p(\tau_1)p(\tau_2)\,d\tau_1\,d\tau_2. \qquad (4.4)$$

In this equation $h_1(t)$ and $h_2(t,\tau)$ are the first- and second-order system response functions (kernels).
By these functions the system is completely characterized.
The non-linearity of the model can be proved by the following deviation:

$$Q(t) = H[p(t)]; \Rightarrow H[p_1(t) + p_2(t)] \neq H(p_1(t)] + H[p_2(t)] \qquad (4.5)$$

$$H[p_1(t)] = Q_1(t) \qquad (4.6)$$

$$= \int_0^t h_1(t-\tau)p_1(\tau)d\tau + \int_0^t\int_0^t h(t-\tau_1,t-\tau_2)p_1(\tau_1)p_1(\tau_2)d\tau_1\,d\tau_2 \quad (4.7)$$

$$H[p_2(t)] = Q_2(t) \qquad (4.8)$$

$$= \int_0^t h_1(t-\tau)p_2(\tau)d\tau + \int_0^t\int_0^t h(t-\tau_1,t-\tau_2)p_2(\tau_1)p_2(\tau_2)d\tau_1\,d\tau_2 \quad (4.9)$$

$$H[p_1(t) + p_2(t)] = Q(t) \qquad (4.10)$$

$$= \int_0^t h_1(t-\tau)(p_1(\tau) + p_2(t))d\tau + \int_0^t\int_0^t h(t-\tau_1,t-\tau_2) \qquad (4.11)$$

$$[p_1(\tau_1) + p_2(\tau_1)][p_1(\tau_2) + p_2(\tau_2)]\,d\tau_1\,d\tau_2.$$

Both kernel functions of the non-linear volterra series model will now be determined. The following manner of solution is proposed by Diskin (1973), and there is also demonstrated the application of the model for non-linearity in the rainfall-runoff system on the watershed.
The kernel functions of the Volterra series model can also be considered system response functions as we have defined them in the case of a linear model. Using the unit impulse function for system input we obtain:

$$p(t) = a\delta(t) \qquad (4.12)$$

$$Q(t) = \int_0^t h_1(\tau)a\delta(t-\tau)d\tau + \int_0^t\int_0^t h_2(\tau_1,\tau_2)a\delta(t-\tau_1)a\delta(t-\tau_2)\,d\tau_1\,d\tau_2 \qquad (4.13)$$

$$Q(t) = ah_1(t) + a^2h_2(t, t) \tag{4.14}$$

$$Q(t)/a = h_1(t) + ah_2(t, t) = HU(t). \tag{4.15}$$

The system function $h_1(t)$ is called the first-order unit response and the function $h_2(t, t)$ the second-order unit response. By use of the common unit hydrograph analysis the quotient $Q(t)/a = HU(t)$ presents a system response of a linear system or model.

Among other things, the system functions of the second-order model have to satisfy the following constraints:

$$\int_0^\infty h_1(t)\,dt = 1 \quad \text{and} \quad \int_0^\infty h_2(t, t)\,dt = 0. \tag{4.16}$$

Considering at least two independent rainfall events with effective rain of volumes of a_1 and a_2 respectively, producing different unit hydrographs $HU_1(t)$ and $HU_2(t)$ in a linear system model, whereby some non-linearity is indicated by different shapes, we may use these functions to determine the first- and second-order system responses $h_1(t)$ and $h_2(t)$ of the non-linear Volterra model.

We obtain a system of two linear equations where the non-linear system functions can be derived as follows:

$$\left. \begin{array}{l} HU_1(t) = h_1(t) + a_1h_2(t, t) \\ HU_2(t) = h_1(t) + a_2h_2(t, t) - 1 \end{array} \right]_+ \quad \left. \begin{array}{l} \cdot \quad a_2 \\ \cdot - a_1 \end{array} \right]_+ \tag{4.17}$$

$$HU_1(t) - HU_2(t) = (a_1 - a_2)h_2(t, t) \tag{4.18}$$

$$h_2(t, t) = [HU_1(t) - HU_2(t)]/(a_1 - a_2) \tag{4.19}$$

$$a_2HU_1(t) - a_1HU_2(t) = (a_2 - a_1)h_1(t) \tag{4.20}$$

$$h_1(t) = [a_2HU_1(t) - a_1HU_2(t)]/(a_2 - a_1). \tag{4.21}$$

To demonstrate a result obtained by this kind of calculation, the basically used unit hydrographs are constrained as follows:

1. $HU_1(t) > HU_2(t)$ if $t < t_s$ (point of intersection)
2. $\max HU_1(t) > \max HU_2(t)$
3. t for $\max HU_1(t) < t$ for $\max HU_2(t)$
4. $a_1/a_2 = 2/1$.

$HU_1(t)$, $HU_2(t)$

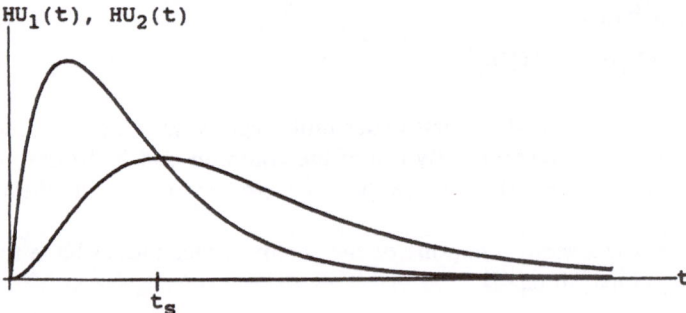

Fig. 4.1 System responses of different input-impulses

$h_2(t,t)$

Fig. 4.2 System response of second order

$h_1(t)$

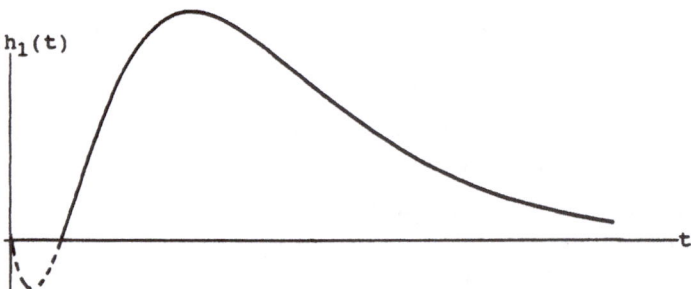

Fig. 4.3 System response of first order

Example to demonstrate the model.
Given: Two rainfall events with a volume of effective rain of $a_1 = 0.6$ mm and $a_2 = 4.94$ mm respectively. Two different unit hydrographs $HU_1(t)$ and $HU_2(t)$ (system responses) of linear system models.

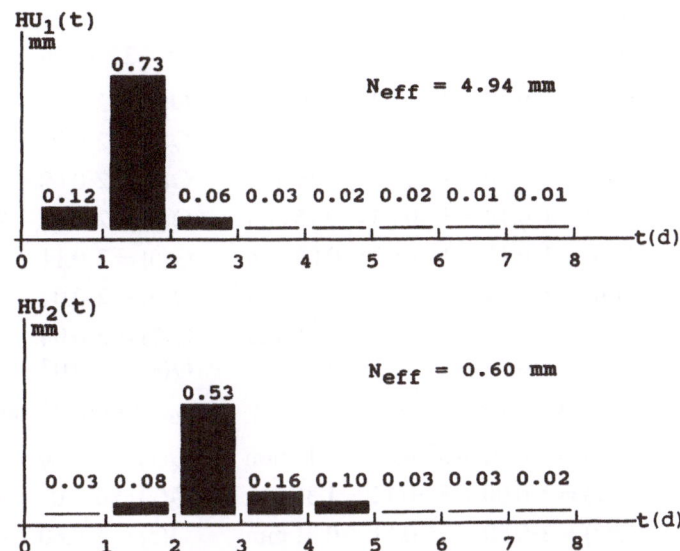

Fig. 4.4 Different unit hydrographs of linear systems

Desired: The first- and second-order system response $h_1(t)$ and $h_2(t)$ of the non-linear Volterra series model.

Solution:

$a_1 = 4.94 \qquad a_2 = 0.60 \qquad 1/(a_1 - a_2) = 0.23$

$h_2(t,t) = [HU_1(t) - HU_2(t)]/(a_1 - a_2)$

$h_2(1,1) = 0.23(0.12 - 0.03) = \qquad 0.02\,\text{mm}$

$h_2(2,2) = 0.23(0.73 - 0.08) = \qquad 0.15\,\text{mm}$

$h_2(3,3) = 0.23(0.06 - 0.53) = \qquad -0.11\,\text{mm}$

$h_2(4,4) = 0.23(0.03 - 0.16) = \qquad -0.03\,\text{mm}$

$h_2(5,5) = 0.23(0.02 - 0.10) = \qquad -0.02\,\text{mm}$

$h_2(6,6) = 0.23(0.02 - 0.03) = \approx -0.01\,\text{mm}$

$h_1(t) = [a_2 HU_1(t) - a_1 HU_2(t)]/(a_2 - a_1)$

$h_1(1) = -0.23(0.6 \cdot 0.12 - 4.94 \cdot 0.03) = \qquad 0.017 = \approx 0.00\,\text{mm}$

$h_1(2) = -0.23(0.6 \cdot 0.73 - 4.94 \cdot 0.08) = -0.009 = \approx 0.00\,\text{mm}$

$h_1(3) = -0.23(0.6 \cdot 0.06 - 4.94 \cdot 0.53) = \qquad 0.593 = \approx 0.60\,\text{mm}$

$h_1(4) = -0.23(0.6 \cdot 0.03 - 4.94 \cdot 0.16) = \qquad 0.177 = \approx 0.18\,\text{mm}$

$h_1(5) = -0.23(0.6 \cdot 0.02 - 4.94 \cdot 0.10) = \qquad 0.111 = \approx 0.11\,\text{mm}$

$h_1(6) = -0.23(0.6 \cdot 0.02 - 4.94 \cdot 0.03) = \qquad 0.031 = \approx 0.04\,\text{mm}$

$h_1(7) = -0.23(0.6 \cdot 0.01 - 4.94 \cdot 0.03) = \qquad 0.032 = \approx 0.04\,\text{mm}$

$h_1(8) = -0.23(0.6 \cdot 0.01 - 4.94 \cdot 0.02) = \qquad 0.035 = \approx 0.03\,\text{mm}.$

The non-linear model will now be applied to simulate runoff curves caused by effective rain with different constant intensities:

$$Q(t) = ah_1(t) + a^2h_2(t, t)$$

1. 1 mm eff. rainfall

$Q(1) = 1 \cdot 0.00 + 1^2 \cdot 0.02 = 0.02\,\text{mm}$

$Q(2) = 1 \cdot 0.00 + 1^2 \cdot 0.15 = 0.15\,\text{mm}$

$Q(3) = 1 \cdot 0.60 - 1^2 \cdot 0.11 = 0.49\,\text{mm}$

$Q(4) = 1 \cdot 0.18 - 1^2 \cdot 0.03 = 0.15\,\text{mm}$

$Q(5) = 1 \cdot 0.11 - 1^2 \cdot 0.02 = 0.09\,\text{mm}$

$Q(6) = 1 \cdot 0.04 - 1^2 \cdot 0.01 = 0.03\,\text{mm}$

$q(7) = 1 \cdot 0.04 \qquad\qquad = 0.04\,\text{mm}$

$Q(8) = 1 \cdot 0.03 \qquad\qquad = 0.03\,\text{mm}$

2. 2 mm eff. rainfall

$Q(1) = 2 \cdot 0.00 + 2^2 \cdot 0.02 = 0.08\,\text{mm}$

$Q(2) = 2 \cdot 0.00 + 2^2 \cdot 0.15 = 0.60\,\text{mm}$

$Q(3) = 2 \cdot 0.60 - 2^2 \cdot 0.11 = 0.76\,\text{mm}$

$Q(4) = 2 \cdot 0.18 - 2^2 \cdot 0.03 = 0.24\,\text{mm}$

$Q(5) = 2 \cdot 0.11 - 2^2 \cdot 0.02 = 0.14\,\text{mm}$

$Q(6) = 2 \cdot 0.04 - 2^2 \cdot 0.01 = 0.04\,\text{mm}$

$Q(7) = 2 \cdot 0.04 \qquad\qquad = 0.08\,\text{mm}$

$Q(8) = 2 \cdot 0.03 \qquad\qquad = 0.06\,\text{mm}.$

3. 3 mm eff. rainfall

$Q(1) = 3 \cdot 0.00 + 3^2 \cdot 0.02 = 0.18\,\text{mm}$

$Q(2) = 3 \cdot 0.00 + 3^2 \cdot 0.15 = 1.35\,\text{mm}$

$Q(3) = 3 \cdot 0.60 - 3^2 \cdot 0.11 = 0.81\,\text{mm}$

$Q(4) = 3 \cdot 0.18 - 3^2 \cdot 0.03 = 0.27\,\text{mm}$

$Q(5) = 3 \cdot 0.11 - 3^2 \cdot 0.02 = 0.15\,\text{mm}$

$Q(6) = 3 \cdot 0.04 - 3^2 \cdot 0.01 = 0.03\,\text{mm}$

$q(7) = 3 \cdot 0.04 \qquad\qquad = 0.12\,\text{mm}$

$Q(8) = 3 \cdot 0.03 \qquad\qquad = 0.09\,\text{mm}$

4. 4 mm eff. rainfall

$Q(1) = 4 \cdot 0.00 + 4^2 \cdot 0.02 = 0.32\,\text{mm}$

$Q(2) = 4 \cdot 0.00 + 4^2 \cdot 0.15 = 2.40\,\text{mm}$

$Q(3) = 4 \cdot 0.60 - 4^2 \cdot 0.11 = 0.64\,\text{mm}$

$Q(4) = 4 \cdot 0.18 - 4^2 \cdot 0.03 = 0.24\,\text{mm}$

$Q(5) = 4 \cdot 0.11 - 4^2 \cdot 0.02 = 0.12\,\text{mm}$

$Q(6) = 4 \cdot 0.04 - 4^2 \cdot 0.01 = 0.00\,\text{mm}$

$Q(7) = 4 \cdot 0.04 \qquad\qquad = 0.16\,\text{mm}$

$Q(8) = 4 \cdot 0.03 \qquad\qquad = 0.12\,\text{mm}.$

This simulation shows that the property of the non-linear model is such that with an increase of effective rainfall the peak of the runoff curve also increases. Moreover, the times to peak decrease if the effective rainfall increases.

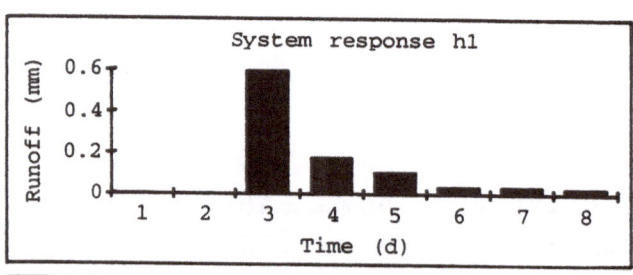

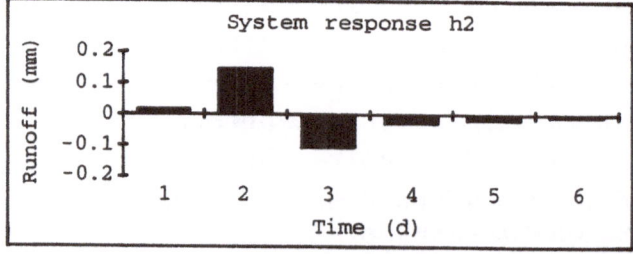

Fig. 4.5 Graphical representation of the system response

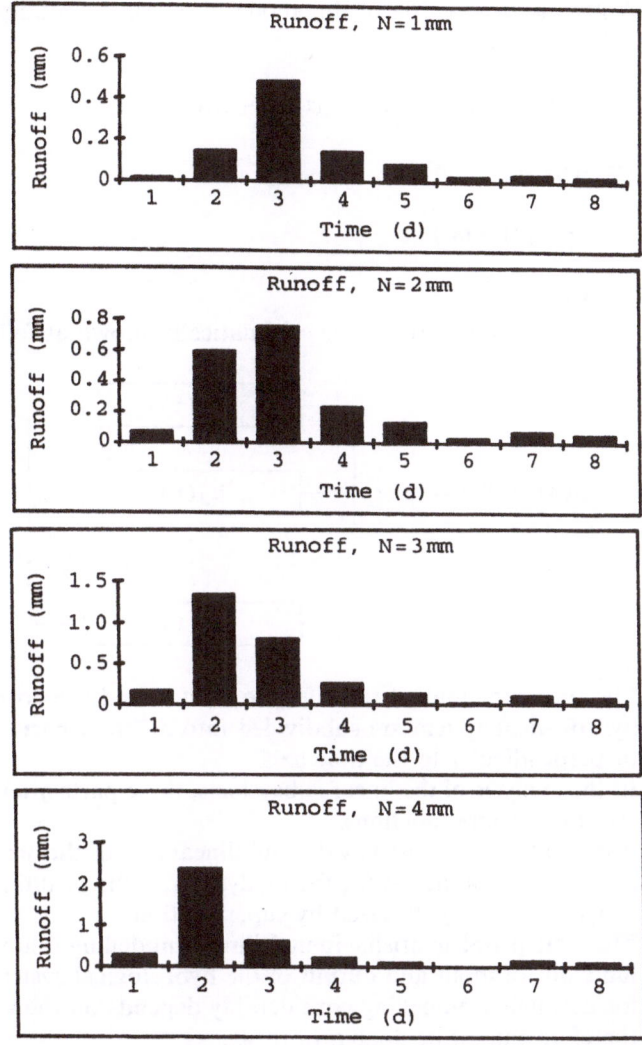

Fig. 4.6 Graphical representation of simulation

4.2   Composition of Parallel Submodels

Linear systems used for modelling of parts derived from a time function for a hydrological system input are called multilinear models. If they are composed parallel to linear submodels, non-linear effects of hydrological systems can be taken into consideration.

A general description of this model type can be mathematically given by the following equation:

$$q(t) = \sum_{i=1}^{n} q_i(t) = \sum_{i=1}^{n} \int_0^t h_i(\tau) p_i(t - \tau) \, d\tau, \tag{4.22}$$

whereby the condition:

$$\sum_{i=1}^{n} p_i(t) = p(t) \tag{4.23}$$

has to be satisfied.

The model structure can be schematically shown as follows:

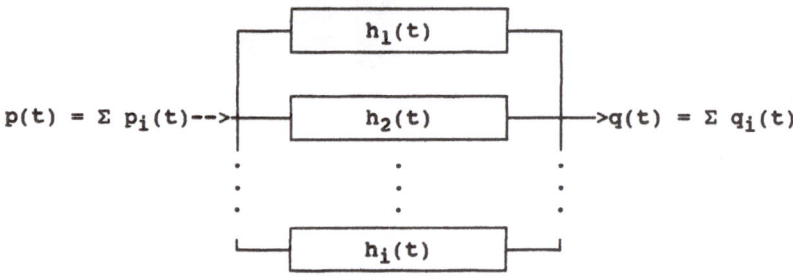

The time functions of the effective rainfall and the corresponding runoff in a hydrological system are subdivided into different parts by lines either parallel or perpendicular to the time axis.

In the analysis of the linear subsystems, these parts are used to determine their system response functions.

For simulation of runoff by the multilinear model, the precipitation is subdivided in the same manner as for the analysis, and the resulting outputs of the linear subsystems are synthesized by superposition.

The main problem arising in multilinear modelling is how to subdivide the time functions of input and output of the hydrological system. Of course, the effect for non-linear modelling considerably depends on the kind of subdivision and therefore cannot be done arbitrarily.

However, practical solutions for subdividing can be found by methods for optimal model adaptation.

Some possibilities for subdividing a discrete rainfall time function are demonstrated below:

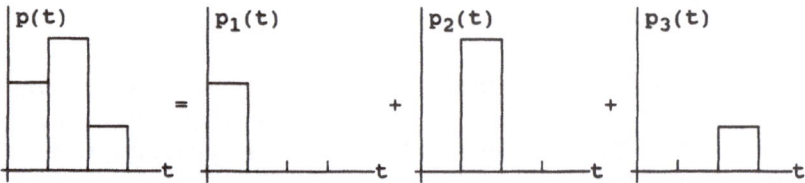

or

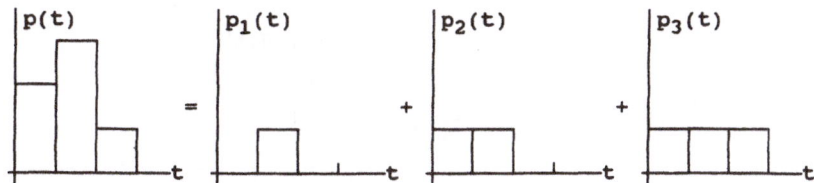

Fig. 4.7 Possible subdividing of a discrete time function

In the following example this method for non-linear modelling is demonstrated:

Given : Time distribution of effective rainfall p(t) and the corresponding runoff curve q(t) in a hydrological system

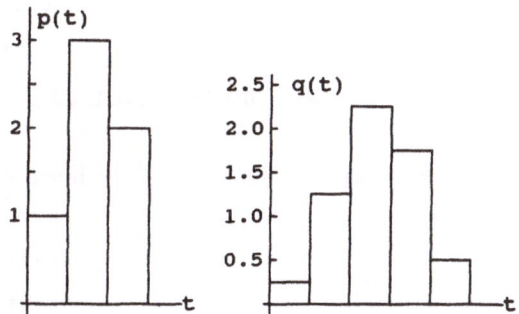

Fig. 4.8 Discrete time function of eff. rain and runoff

Desired : System response functions $h_1(t)$ and $h_2(t)$ for non-linear modelling (analysis).
Run-off simulation for comparing with an only linear modelling (synthesis)

Solution:

a) Analysis (system identification):

The subdivision has been made by drawing a line parallel to the time axis at the value of 1 mm of rain.

The amounts of the precipitation parts obtained have to be equal to those of the corresponding runoff.

The following two parallel linear submodels are then formed:

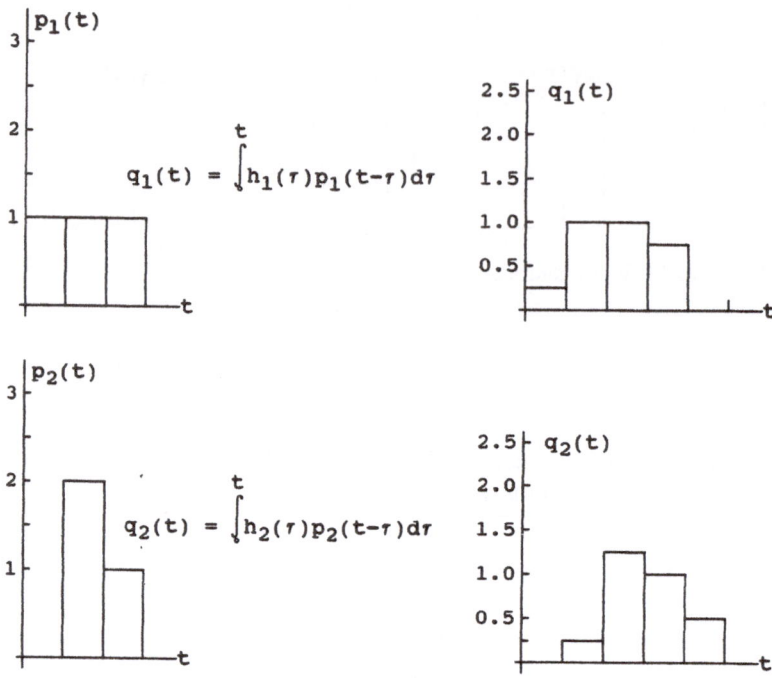

Fig. 4.9 Input and output function of two linear submodels

The system response functions of the linear submodels are:

Fig. 4.10 System response functions of the two linear submodels

The system response function of only one linear system is:

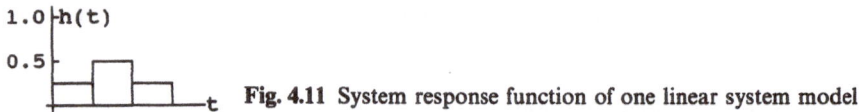   **Fig. 4.11** System response function of one linear system model

b) Synthesis (simulation):
For simulation a runoff curve by non-linear modelling as well as by only linear modelling in comparison an effective rainfall with a constant intensity of 2 mm is assumed.

In the case of non-linear modelling there is the following subdivision:

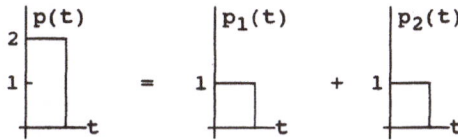

Fig. 4.12 Rainfall subdivision for non-linear modelling

The non-linear model via two parallel submodels results in the following simulated runoff by superposition:

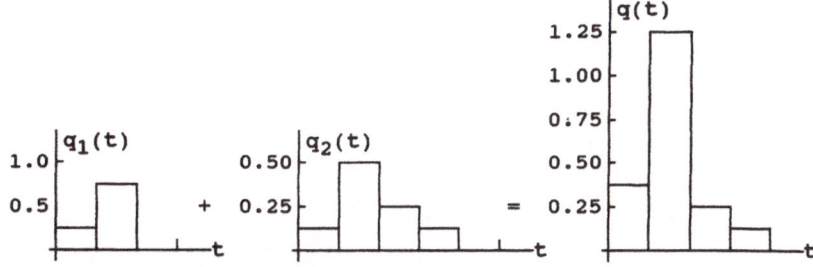

Fig. 4.13 Superposition of simulated discrete runoff functions

By comparison, the linear model simulates runoff by a time distribution without any effects of non-linearity as shown below.

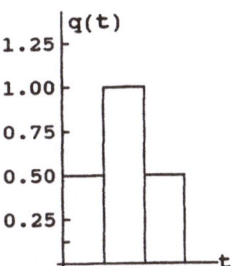

Fig. 4.14 Simulated runoff function by linear modelling

In the curves (histograms) for the runoff simulation there are obviously considerable differences in the shapes, indicating a wide range of non-linear effects.

Consequently, in non-linear catchment behaviour the composition of parallel linear submodels appears to be a powerful tool for modelling.

Bibliography

Amorocho J, Orlob G T (1961) Nonlinear analysis of hydrologic systems. Water Resources Center, Nr. 40 University of California, Berkeley

Beyer O, Girlich H J Zschiesche H U, (1978) Stochastische Prozesse. Verlag Harri Deutsch, Thun, Frankfurt/Main

Blank D, Delleur W, Giorgini A (1971) Oscillatory kernel functions in linear hydrological models. Water Resources Res. 7(5)

Boneh A (1971) Identification of the kernel functions of hydrologic systems from independent multi-storm records. D. Sc. Thesis Technion – Israel Institute of Technology Haifa, Israel

Bruen M, Dooge J C, I (1984) An efficient and robust method for estimating unit hydrograph ordinates. J. Hydrol. 70

Clark C O (1945) Storage and the unit hydrograph. Trans. ASCE. 110:1419–1446

Diskin M H (1973) Proceedings of the Second International Symposium in Hydrology Floods and Drought, Fort Collins, Colorado 11–13 Sept 1973

Diskin M H, Boneh, A, Golan A (1972) Properties of the kernels for time invariant, initially relaxed, second order, surface runoff systems. J. Hydrol. 17 (1)

Dyck G, Angewandte Hydrologie, Teil 1, 2. (1977) Wilhelm Ernst & Sohn, Berlin, Mümchen

Gois R S S (1977) Desenvolvimento do hidrograma unitario da bacia do Rio Mamuaba. Universidade Federal da Paraiba, Centro de Ciencias e Technologia, Departamento de Engenaria Civil, Tese, Campina Grande

Graupe D (1972) Identification of systems. Van Nostrand Reinhold Co, New York, Cincinnati

Irving J, Mullineux N (1967) Mathematics in physics and engineering. Academic Press New York, London

Jacoby S L S (1966) A mathematical model for non-linear hydrologic systems, J. Geophys. Res. 71:20

Karman T, Biot M A (1948) Metodos matematicos em engenharia. Livraria Kosmos Editora, Erich Eichner & CIA. LTDA., Rio de Janeiro

Kavvas M L (1973) Removal of unit hydrograph oscillations by filtering. Proceedings of the Second International Symposium in Hydrology. Floods and Droughts. Fort Collins, Colorado, 11.–13. Sept.

Koehler G (1976) Niederschlags – Abfluß – Modelle für kleine Einzugsgebiete. Schriftenreihe des Kuratoriums für Wasser und Kulturbauwesen Heft 25 Paul Parey, Hamburg

Lampard D G (1955) The response of linear network to suddenly applied stationary noise, IRE Trans. Circuit theory. Vol. CT-2, March 1955

Linsley R K, Kohler M A, Paulhus J L H (1975) Hydrology for engineers, McGraw-Hill New York

Ludwig K (1979) Hydrologische Verfahren und Beispiele für die wasserwirtschaftliche Bemessung von Hochwasserrückhaltebecken. Schriftenreihe des Kuratoriums für Wasser- und Kulturbauwesen Heft 44 Paul Parey, Hamburg

Minshall N E (1986) Predicting storm runoff on small experimental watersheds, J. Hydraul. Div. Proc. ASCE HY8 Vol. 86

Murray R (1977) Spiegel, Laplace-Transformationen. McGraw-Hill, New York

Nelson L. de Sousa Pinto (1973) Hidrologia de superficie, Edgard Blücher Ltda, Sao Paulo (SP), Brazil

Neuman S P, Marsily G (1976) Identification of linear system response by parametric programming, Water Resources Res. Vol. 12 (2)

Papoulis A, (1984) Probability, random variables, and stochastic processes, McGraw-Hill, New York

Plate E J, Schultz G A, Seus J G, Wittenberg H (1977) Ablauf von Hochwasserwellen in Gerinnen, Schriftenreihe des Kuratoriums für Wasser- und Kulturbauwesen. Heft 27 Paul Parey, Hamburg

Raudkivi A J (1979) Hydrology, Pergamon Press, New York

Schlitt H (1960) Systemtheorie für regellose Vorgänge, Springer, Berlin, Heidelberg, New York

Schultz G A (1968) Bestimmung theoretischer Abflußganglinien durch elektronische Berechnung von Niederschlagskonzentration und Retension (Hyreun-Verf.). Versuchsanstalt für Wasserbau der TH München, Oskar V. Miller-Institut, Bericht Nr. 11

Schwarz J R, Friedland B (1965) Linear Systems. McGraw-Hill, New York

Viessmann W. Knapp J W, Lewis G L, Harbaugh T E (1977) Introduction to hydrology. Harper & Row, New York

Villela S M, Mattos A (1975) Hidrologia aplicada. McGraw-Hill Sao Paulo (SP), Brazil

```
90 KEY OFF
100 REM              ******************************************
110 REM              *          PROGRAM    'PMSB1'            *
120 REM              *     WRITTEN BY DR.A./R.A. LATTERMANN   *
130 REM              *                1987                    *
140 REM              ******************************************
150 REM
160 CLS:BEEP
162 REM
163 REM UTILIZATION OF THE PROGRAM AND DATA INPUT
165 REM
170 GOSUB 2000
180 PRINT:PRINT
190 PRINT TAB(10)"THE PROGRAM 'PMSB1' CALCULATES THE MEAN AREAL
PRECIPITATION"
200 PRINT TAB(10)"BY USE OF THE I S O H Y E T A L METHOD"
210 PRINT
220 PRINT TAB(10)"DATA OF INPUT ARE:"
230 PRINT
240 PRINT TAB(10)"1.      MAX. NUMBER OF TIME INTERVALS OF THE
CONSIDERED"
250 PRINT TAB(10)"        PRECIPITATION EVENT              NI=";
:INPUT NI
260 PRINT TAB(10)"2.      MAX. NUMBER OF PARTIAL AREAS   AI=";
:INPUT PI
270 PRINT TAB(10)"3.      AVERAGE ISOHYETALS OF PARTIAL AREAS
PER TIMES"
280 A=NI:IF NI<PI THEN A=PI
290 DIM P(A,A),PM(A),PP(A),ISO(A,A),SUP(A)
300 FOR I=1 TO NI:LOCATE 16,10:PRINT "TIME INTERVAL "; I
:LOCATE 17,10:
305 PRINT "NUMBER OF PARTIAL AREAS                     AI = ";
:INPUT PP(I) :FOR J=1 TO PP(I)
310 LOCATE 19,10:PRINT "SIZE OF THE PARTIAL AREA
"; J; "= "; :INPUT P(I,J)
315 LOCATE 20,10:PRINT "AVERAGE ISOHYETAL PER PARTIAL AREA
(MM)"; J; "= "; :INPUT ISO(I,J): NEXT J,I
317 FOR I=1 TO NI
318 SUP(I)=0:FOR J=1 TO PP(I):SUP(I)=SUP(I)+P(I,J):NEXT J
319 NEXT I
320 REM
321 REM CALCULATION ..
322 REM
325 FOR I=1 TO NI
330 PM(I)=0
340 FOR J=1 TO PP(I)
350 PM(I)=PM(I)+P(I,J)*ISO(I,J)
360 NEXT J
370 PM(I)=PM(I)/SUP(I)
380 NEXT I
```

```
390 CLS:BEEP
400 GOSUB 2000
401 REM
402 REM FOR DISPLAY  ..
403 REM
410 PRINT:PRINT
420 PRINT TAB(10)"MEAN AREAL PRECIPITATION (INTERVAL)
(MM):":PRINT
430 FOR I=1 TO NI
440 PRINT TAB(10)"PM("; I; ")    =    "; :PRINT USING "###.###";
PM(I):NEXT I
450 PRINT
460 PRINT TAB(10)"==== PRINTING OF THE RESULT (Y/N) ???"
470 U$=INKEY$
480 IF U$="" THEN 470
490 IF U$="Y" OR U$="y" THEN GOSUB 3000
495 GOTO 12000
500 CLS
510 GOSUB 2000
520 PRINT:PRINT
530 PRINT TAB(10)"========== END OF THE PROGRAM =========="
540 KEY ON:PLAY "O4T200MLL8GAGAL2C":END
2000 PRINT:PRINT TAB(10)"[ PROGRAM :    P  M  S  B  1  ]
"; DATE$:PRINT:RETURN
3000 REM
3001 REM FOR PRINTER ..
3002 REM
3005 LPRINT:LPRINT
3010 LPRINT CHR$(14); "            P  M  S  B  1";
CHR$(20):LPRINT
3020 LPRINT CHR$(15); "
WRITTEN BY DR.A./R.A. LATTERMANN"; CHR$(18)
3030 LPRINT:LPRINT:LPRINT
3038 LPRINT TAB(10)"
"; DATE$:LPRINT
3040 LPRINT TAB(10)"THE PROGRAM 'PMSB1' CALCULATES THE MEAN
AREAL PRECIPITATION"
3050 LPRINT TAB(10)"BY USE OF THE I S O H Y E T A L  METHOD"
3060 LPRINT:LPRINT
3070 LPRINT TAB(10)"DATA OF INPUT ARE:":LPRINT
3080 LPRINT TAB(10)"1.   MAX. NUMBER OF TIME INTERVALS OF THE
CONSIDERED"
3090 LPRINT TAB(10)"      PRECIPITATION EVENT             NI = ";
NI:LPRINT
3100 LPRINT TAB(10)"2.   MAX. NUMBER OF PARTIAL AREAS  AI = ";
PI:LPRINT
3110 LPRINT TAB(10)"3.   AVERAGE ISOHYETALS OF PARTIAL AREAS
PER TIMES"
3120 FOR I=1 TO NI:LPRINT:LPRINT TAB(10)"TIME INTERVAL :"; I
3125 LPRINT TAB(10)"NUMBER OF PARTIAL AREAS  AI:"; PP(I)
3130 LPRINT TAB(10)"INTERVAL:"; TAB(27)"AVERAGE ISOHYETALS
(MM):"; TAB(45)"SIZE OF THE PARTIAL AREA (AU):"
3140 FOR J=1 TO PP(I)
```

```
3150 LPRINT TAB(13)J; TAB(35)ISO(I,J); TAB(50)P(I,J)
3160 NEXT J,I:LPRINT
3170 LPRINT:LPRINT:LPRINT
3180 LPRINT TAB(10)"RESULTS :":LPRINT
3190 LPRINT:LPRINT TAB(10)"MEAN AREAL PRECIPITATION (INTERVAL)
:":LPRINT
3200 LPRINT TAB(10)"INTERVAL:"; TAB(45)"MEAN AREAL
PRECIPIT.(MM):"
3210 FOR I=1 TO NI
3220 LPRINT TAB(13); I; TAB(50); PM(I)
3230 NEXT I
3240 RETURN
12000 REM
12002 REM END ..
12005 REM
12009 PRINT
12010 PRINT TAB(10)"==== (N)EW CALCULATION OR (E)ND ??? ===="
12020 U$=INKEY$
12030 IF U$="" THEN 12020
12040 IF U$="N" OR U$="n" THEN RUN
12050 GOTO 500
12060 END
```

```
90 KEY OFF
100 REM          *****************************************
110 REM          *          PROGRAM   'PMSB2'            *
120 REM          *     WRITTEN BY DR.A./R.A. LATTERMANN  *
130 REM          *               1987                    *
140 REM          *****************************************
150 REM
160 CLS:BEEP
162 REM
163 REM UTILIZATION OF THE PROGRAM AND DATA INPUT
165 REM
170 GOSUB 2000
180 PRINT:PRINT
190 PRINT TAB(10)"THE PROGRAM 'PMSB2' CALCULATES THE MEAN AREAL
PRECIPITATION"
200 PRINT TAB(10)"BY USE OF THE T H I E S S E N - METHOD"
210 PRINT
220 PRINT TAB(10)"DATA OF INPUT ARE:"
230 PRINT
240 PRINT TAB(10)"1.     MAX. NUMBER OF TIME INTERVALS OF THE
CONSIDERED"
250 PRINT TAB(10)"          PRECIPITATION EVENT            NI=";
:INPUT NI
260 PRINT TAB(10)"2.     MAX. NUMBER OF GAGE-STATIONS  PI=";
:INPUT PI
270 PRINT TAB(10)"3.     MEASURED PRECIPITATION VALUES:"
280 A=NI:IF NI<PI THEN A=PI
290 DIM P(A,A),PM(A)
300 FOR I=1 TO PI:LOCATE 16,10:PRINT "STATION "; I:FOR J=1 TO
NI
310 LOCATE 17,10:PRINT "MEASURED PRERCIPITATION (MM) OF
INTERVAL  "; J; " =                 "; :LOCATE 17,55:INPUT
P(I,J):NEXT J,I
312 PRINT:PRINT TAB(10)"4.     AREA OF T H I E S S E N -
POLYGONAL :"
314 FOR I=1 TO PI:LOCATE 21,10:PRINT "STATION "; I
316 LOCATE 22,10:PRINT "THIESSEN-AREA:             "; LOCATE
22,55:INPUT TH(I):NEXT I
318 TS=0:FOR I=1 TO PI:TS=TS+TH(I):NEXT I
320 REM
321 REM CALCULATION ..
322 REM
325 FOR J=1 TO NI
330 PM(J)=0
340 FOR I=1 TO PI
350 PM(J)=PM(J)+P(I,J)*TH(I)
360 NEXT I
370 PM(J)=PM(J)/TS
380 NEXT J
390 CLS:BEEP
```

```
400 GOSUB 2000
401 REM
402 REM FOR DISPLAY ..
403 REM
410 PRINT:PRINT
420 PRINT TAB(10)"MEAN AREAL PRECIPITATION (INTERVAL)
(MM):":PRINT
430 FOR I=1 TO NI
440 PRINT TAB(10)"PM("; I; ")   =   "; :PRINT USING "###.###";
PM(I):NEXT I
450 PRINT
460 PRINT TAB(10)" PRINTING OF THE RESULT (Y/N) ???"
470 U$=INKEY$
480 IF U$="" THEN 470
490 IF U$="Y" OR U$="y" THEN GOSUB 3000
495 GOTO 12000
500 CLS
510 GOSUB 2000
520 PRINT:PRINT
530 PRINT TAB(10)"========== END OF THE PROGRAM =========="
540 KEY ON:PLAY "O4T200MLL8GAGAL2C":END
2000 PRINT:PRINT TAB(10)"[ PROGRAM :    P  M  S  B  2  ]
"; DATE$:PRINT:RETURN
3000 REM
3001 REM FOR PRINTING  ..
3002 REM
3005 LPRINT:LPRINT
3010 LPRINT CHR$(14); "            P  M  S  B  2";
CHR$(20):LPRINT
3020 LPRINT CHR$(15); "
WRITTEN BY DR.A./R.A. LATTERMANN"; CHR$(18)
3030 LPRINT:LPRINT:LPRINT
3038 LPRINT TAB(10)"
"; DATE$:LPRINT
3040 LPRINT TAB(10)"THE PROGRAM 'PMSB2' CALCULATES THE MEAN
AREAL PRECIPITATION"
3050 LPRINT TAB(10)"BY USE OF THE T H I E S S E N -METHOD"
3060 LPRINT:LPRINT
3070 LPRINT TAB(10)"DATA OF INPUT ARE:":LPRINT
3080 LPRINT TAB(10)"1.   MAX. NUMBER OF TIME INTERVALS OF THE
CONSIDERED"
3090 LPRINT TAB(10)"    PRECIPITATION EVENT              NI = ";
NI:LPRINT
3100 LPRINT TAB(10)"2.   MAX. NUMBER OF GAGE STATIONS  PI = ";
PI:LPRINT
3110 LPRINT TAB(10)"3.   MEASURED PRECIPITATION VALUES"
3120 FOR I=1 TO PI:LPRINT:LPRINT TAB(10)"STATION :"; I
3130 LPRINT TAB(10)"INTERVAL:"; TAB(45)"MEASURED PRECIPITATION
(MM):"
3140 FOR J=1 TO NI
3150 LPRINT TAB(13)J; TAB(50)P(I,J)
3160 NEXT J,I:LPRINT
```

```
3162 LPRINT TAB(10)"4.    AREAS OF THE T H I E S S E N -
POLYGONALS:"
3164 LPRINT:LPRINT TAB(10)"STATION:"; TAB(45)"T H I E S S E N -
AREA:"
3166 FOR I=1 TO PI
3167 LPRINT TAB(13)I; TAB(50)TH(I)
3168 NEXT I
3170 LPRINT:LPRINT:LPRINT
3180 LPRINT TAB(10)"RESULTS :":LPRINT
3190  LPRINT:LPRINT TAB(10)"MEAN AREAL PRECIPITATION (INTERVAL)
:":LPRINT
3200 LPRINT TAB(10)"INTERVAL:"; TAB(45)"MEAN AREAL
PRECIPIT.(MM):"
3210 FOR I=1 TO NI
3220 LPRINT TAB(13); I; TAB(50); PM(I)
3230 NEXT I
3240 RETURN
12000 REM
12002 REM END ..
12005 REM
12009 PRINT
12010 PRINT TAB(10)"==== (N)EW CALCULATION OR (E)ND ??? ===="
12020 U$=INKEY$
12030 IF U$="" THEN 12020
12040 IF U$="N" OR U$="n" THEN RUN
12050 GOTO 500
12060 END
```

```
90 KEY OFF
100 REM        *****************************************
110 REM        *           PROGRAM   'PMSB3'           *
120 REM        *     WRITTEN BY DR.A./R.A. LATTERMANN  *
130 REM        *              1987                     *
140 REM        *****************************************
150 REM
160 CLS:BEEP
162 REM
164 REM UTILISATION OF THE PROGRAM AND DATA INPUT
166 REM
170 GOSUB 2000
180 PRINT:PRINT
190 PRINT TAB(10)"THE PROGRAM 'PMSB3' CALCULATES THE MEAN AREAL
PRECIPITATION"
200 PRINT TAB(10)"BY USE OF THE METHOD OF ARITHMETIC MEAN"
210 PRINT:PRINT
220 PRINT TAB(10)"DATA OF INPUT ARE: "
230 PRINT
240 PRINT TAB(10)"1.     MAX. NUMBER OF TIME INTERVALS OF THE
CONSIDERED"
250 PRINT TAB(10)"      PRECIPITATION EVENT              NI=";
:INPUT NI:PRINT
260 PRINT TAB(10)"2.     MAX. NUMBER OF GAGE STATIONS  PI=";
:INPUT PI:PRINT
270 PRINT TAB(10)"3.     MEASURED PRECIPITATION VALUES:":PRINT
280 A=NI:IF NI<PI THEN A=PI
290 DIM P(A,A),PM(A)
300 FOR I=1 TO PI:LOCATE 19,10:PRINT "STATION "; I:FOR J=1 TO
NI
310 LOCATE 21,10:PRINT "MEASURED PRECIPITATION OF INTERVAL (MM)
"; J; " =              ";:LOCATE 21,55:INPUT P(I,J):NEXT J,I
315 REM
316 REM CALCULATION    ..
317 REM
320 FOR J=1 TO NI
330 PM(J)=0
340 FOR I=1 TO PI
350 PM(J)=PM(J)+P(I,J)
360 NEXT I
370 PM(J)=PM(J)/PI
380 NEXT J
390 CLS:BEEP
391 REM
392 REM FOR DISPLAY   ..
393 REM
400 GOSUB 2000
410 PRINT:PRINT
420 PRINT TAB(10)"MEAN AREAL PRECIPITATION (INTERVAL)
(MM):":PRINT
```

```
430 FOR I=1 TO NI
440 PRINT TAB(10)"PM("; I; ")    =    "; :PRINT USING "###.###";
PM(I):NEXT I
450 PRINT
460 PRINT TAB(10)"==== PRINTING OF THE RESULT (Y/N) ???"
470 U$=INKEY$
480 IF U$="" THEN 470
490 IF U$="Y" OR U$="y" THEN GOSUB 3000
495 GOTO 12000
500 CLS
510 GOSUB 2000
520 PRINT:PRINT
530 PRINT TAB(10)"========== END OF THE PROGRAM =========="
540 KEY ON:PLAY "O4T200L8MLGAGAL4C":END
2000 PRINT:PRINT TAB(10)"[ PROGRAM :    P   M   S   B   3   ]
"; DATE$:PRINT:RETURN
3000 REM
3001 REM FOR PRINTER ..
3003 REM
3009 LPRINT:LPRINT
3010 LPRINT CHR$(14); "              P   M   S   B   3";
CHR$(20):LPRINT
3020 LPRINT CHR$(15); "
WRITTEN BY DR.A./R.A. LATTERMANN"; CHR$(18)
3030 LPRINT:LPRINT:LPRINT
3038 LPRINT TAB(10)"
"; DATE$:LPRINT
3040 LPRINT TAB(10)"THE PROGRAM 'PMSB3' CALCULATES THE MEAN
AREAL PRECIPITATION"
3050 LPRINT TAB(10)"BY USE OF THE METHOD OF ARITHMETIC MEAN"
3060 LPRINT:LPRINT
3070 LPRINT TAB(10)"DATA OF INPUT ARE:":LPRINT
3080 LPRINT TAB(10)"1.   MAX. NUMBER OF TIME INTERVALS OF THE
CONSIDERED"
3090 LPRINT TAB(10)"       PRECIPITATION EVENT            NI = ";
NI:LPRINT
3100 LPRINT TAB(10)"2.   MAX. NUMBER OF GAGE STATIONS  PI = ";
PI:LPRINT
3110 LPRINT TAB(10)"3.   MEASURED PRECIPITATION VALUES:"
3120 FOR I=1 TO PI:LPRINT:LPRINT TAB(10)"STATION :"; I
3130 LPRINT TAB(10)"INTERVAL:"; TAB(45)"MEASURED PRECIPITATION
(MM):"
3140 FOR J=1 TO NI
3150 LPRINT TAB(13)J; TAB(50)P(I,J)
3160 NEXT J,I
3170 LPRINT:LPRINT:LPRINT
3180 LPRINT TAB(10)"RESULTS :":LPRINT
3190 LPRINT:LPRINT TAB(10)"MEAN AREAL PRECIPITATION (INTERVAL)
:":LPRINT
3200 LPRINT TAB(10)"INTERVAL:"; TAB(45)"MEAN AREAL
PRECIPIT.(MM):"
3210 FOR I=1 TO NI
3220 LPRINT TAB(13); I; TAB(50); PM(I)
```

```
3230 NEXT I
3240 RETURN
12000 REM
12001 REM END ..
12002 REM
12005 PRINT
12010 PRINT TAB(10)"==== (N)EW CALCULATION OR (E)ND ?? ===="
12020 U$=INKEY$
12030 IF U$="" THEN 12020
12040 IF U$="N" OR U$="n" THEN RUN
12050 GOTO 500
12060 END
```

```
90 KEY OFF
100 REM              ****************************************
110 REM              *            PROGRAM   'PMSB4'         *
120 REM              *     WRITTEN BY DR.A./R.A. LATTERMANN *
130 REM              *                 1987                 *
140 REM              ****************************************
150 REM
160 CLS:BEEP
161 REM
162 REM UTILIZATION OF THE PROGRAM AND DATA INPUT
163 REM
170 GOSUB 2000
180 PRINT:PRINT
190 PRINT TAB(10)"THE PROGRAM 'PMSB4' CALCULATES THE MEAN AREAL
PRECIPITATION"
200 PRINT TAB(10)"BY USE OF THE G R I D - P O I N T  METHOD"
210 PRINT:PRINT
220 PRINT TAB(10)"DATA OF INPUT ARE:"
230 PRINT
240 PRINT TAB(10)"1.     MAX. NUMBER OF TIME INTERVALS OF THE
CONSIDERED"
250 PRINT TAB(10)"        PRECIPITATION EVENT              NI=";
:INPUT NI:PRINT
260 PRINT TAB(10)"2.     MAX. NUMBER OF GAGE STATIONS  PI=";
:INPUT PI:PRINT
270 PRINT TAB(10)"3.      MEASURED PRECIPITATION VALUES:":PRINT
280 A=NI:IF NI<PI THEN A=PI
290 DIM P(500,A),PM(A)
300 FOR I=1 TO PI:LOCATE 19,10:PRINT "STATION "; I:FOR J=1 TO
NI
310 LOCATE 21,10:PRINT "MEASURED PRECIPITATION (MM) OF INTERVAL
"; J; " =             ";:LOCATE 21,55:INPUT P(I,J):NEXT J,I
312 REM
313 REM CALCULATION ..
314 REM
315 GOSUB 30000:DIM QS(QUI),V(QUI,4),H(QUI),HN(A)
320 FOR I=1 TO QUI:QS(I)=0:FOR J=1 TO 4:IF QU(I,J)=0 THEN 340
330 QS(I)=QS(I)+1/(QU(I,J)^2)
340 NEXT J,I
350 FOR I=1 TO QUI:FOR J=1 TO 4:V(I,J)=0:IF QU(I,J)=0 THEN 370
360 V(I,J)=(1/QU(I,J)^2)/QS(I)
370 NEXT J,I
380 FOR K=1 TO NI:FOR I=1 TO QUI:H(I)=0:FOR J=1 TO 4:IF
QP(I,J)=0 THEN P(QP(I,J),K)=0
385 H(I)=H(I)+V(I,J)*P(QP(I,J),K):NEXT J,I
386 HN(K)=0:FOR I=1 TO QUI:HN(K)=HN(K)+H(I)/QUI:NEXT I,K
390 CLS:BEEP
392 REM
393 REM DISPLAY    ..
394 REM
```

```
400 GOSUB 2000
410 PRINT:PRINT
420 PRINT TAB(10)"MEAN AREAL PRECIPITTATION (INTERVAL)
(MM):":PRINT
430 FOR I=1 TO NI
440 PRINT TAB(10)"PM("; I; ")    =    "; :PRINT USING "###.###";
HN(I):NEXT I
450 PRINT
460 PRINT TAB(10)"==== PRINTING OF THE RESULT (Y/N) ???"
470 U$=INKEY$
480 IF U$="" THEN 470
490 IF U$="Y" OR U$="y" THEN GOSUB 3000
495 GOTO 12000
500 CLS
510 GOSUB 2000
520 PRINT:PRINT
530 PRINT TAB(10)"=========== END OF THE PROGRAM =========="
540 KEY ON:PLAY "O4T200L8MLGAGAL4C":END
2000 PRINT:PRINT TAB(10)"[ PROGRAM :    P  M  S  B  4  ]
"; DATE$:PRINT:RETURN
3000 REM
3001 REM PRINTER ..
3002 REM
3009 LPRINT:LPRINT
3010 LPRINT CHR$(14); "              P  M  S  B  4";
CHR$(20):LPRINT
3020 LPRINT CHR$(15); "
WRITTEN BY DR.A./R.A. LATTERMANN"; CHR$(18)
3030 LPRINT:LPRINT:LPRINT
3038 LPRINT TAB(10)"
"; DATE$:LPRINT
3040 LPRINT TAB(10)"THE PROGRAM 'PMSB4' CALCULATES THE MEAN
AREAL PRECIPITATION"
3050 LPRINT TAB(10)"BY USE OF THE GRID-POINT METHOD"
3060 LPRINT:LPRINT
3070 LPRINT TAB(10)"DATA OF INPUT ARE:":LPRINT
3080 LPRINT TAB(10)"1.    MAX. NUMBER OF TIME INTERVALS OF THE
CONSIDERED"
3090 LPRINT TAB(10)"        PRECIPITATION EVENT                NI = ";
NI:LPRINT
3100 LPRINT TAB(10)"2.    MAX. NUMBER OF GAGE STATIONS  PI = ";
PI:LPRINT
3110 LPRINT TAB(10)"3.    MEASURED PRECIPITATION VALUES:"
3120 FOR I=1 TO PI:LPRINT:LPRINT TAB(10)"STATION :"; I
3130 LPRINT TAB(10)"INTERVAL:"; TAB(45)"MEASURED PRECIPITATION
(MM):"
3140 FOR J=1 TO NI
3150 LPRINT TAB(13)J; TAB(50)P(I,J)
3160 NEXT J,I
3165 GOSUB 40000
3170 LPRINT:LPRINT:LPRINT
3180 LPRINT TAB(10)"RESULTS:":LPRINT
```

```
3190 LPRINT:LPRINT TAB(10)"MEAN AREAL PRECIPITATION (INTERVAL)
:":LPRINT
3200 LPRINT TAB(10)"INTERVAL:"; TAB(45)"MEASURED PRECIPITATION
(MM):"
3210 FOR I=1 TO NI
3220 LPRINT TAB(13); I; TAB(50); HN(I)
3230 NEXT I
3240 RETURN
12000 REM
12001 REM END ..
12002 REM
12009 PRINT
12010 PRINT TAB(10)"==== (N)EW CALCULATION OR (E)ND ??? ===="
12020 U$=INKEY$
12030 IF U$="" THEN 12020
12040 IF U$="N" OR U$="n" THEN RUN
12050 GOTO 500
12060 END
30000 REM
30001 REM GRID-POINT INPUT ..
30002 REM
30009 PRINT
30010 FOR KL=11 TO 21:LOCATE KL,10:PRINT SPACE$(70):NEXT KL
30015 GOSUB 45000:IF LOT=1 THEN RETURN
30017 FOR KL=11 TO 21:LOCATE KL,10:PRINT SPACE$(70):NEXT KL
30020 LOCATE 12,10:PRINT "4.        GRID-POINT VALUES:"
30030 LOCATE 13,10:PRINT "         MAX. NUMBER OF GRID-
POINTS:"; :INPUT QUI
30040 DIM QU(QUI,4),QP(QUI,4)
30050 LOCATE 15,10:PRINT "         DISTANCES BETWEEN THE GRID-
POINT CENTERS" :LOCATE 16,10:PRINT "        AND THE PROXIMATE
STATIONS/QUADRANT :"
30060 FOR I=1 TO QUI:FOR J=1 TO 4
30070 LOCATE 18,10:PRINT "         GRID-POINT"; I; ",
QUADRANT"; J; ", STATION-NR." ; :INPUT QP(I,J)
30080 LOCATE 19,10:PRINT "         DISTANCES   :"; :INPUT
QU(I,J):LOCATE 18,10:PRINT SPACE$(60):LOCATE 19,10:PRINT
SPACE$(60):NEXT J,I
30090 LOCATE 20,10:PRINT "SAVING THE GRID-POINT VALUES (Y/N) ?"
30100 U$=INKEY$
30110 IF U$="" THEN 30100
30120 IF U$="Y" OR U$="y" THEN 30140
30130 RETURN
30140 LOCATE 20,10:PRINT SPACE$(60):LOCATE 20,10:PRINT "NAME OF
THE FILE:" ; :INPUT NN$
30150 OPEN NN$ FOR OUTPUT AS #1:PRINT #1,QUI
30160 FOR I=1 TO QUI:FOR J=1 TO 4:PRINT #1,QP(I,J):PRINT
#1,QU(I,J):NEXT J,I
30170 CLOSE
30180 RETURN
40000 LPRINT
40020 LPRINT TAB(10)"4.        GRID-POINT VALUES :"
```

```
40030 LPRINT TAB(10)"            MAX. NUMBER OF GRID-POINT VALUES
:"; QUI
40050 LPRINT TAB(10)"            DISTANCES BETWEEN THE GRID-POINT
CENTERS " :LPRINT TAB(10)"            AND THE PROXIMATE
STATIONS/QUADRANT :"
40060 LPRINT TAB(10)"GRID-POINT"; TAB(25)"QUADRANT";
TAB(40)"STATION-NR." ; TAB(60)"DISTANCES"
40070 FOR I=1 TO QUI:FOR J=1 TO 4
40080 LPRINT TAB(15)I; TAB(31)J; TAB(48)QP(I,J);
TAB(65)QU(I,J):NEXT J,I
40090 RETURN
45000 LOCATE 15,10:PRINT "READING THE GRID-POINT VALUES FROM A
FILE (Y/N)?"
45010 U$=INKEY$
45020 IF U$="" THEN 45010
45030 IF U$="Y" OR U$="y" THEN 45050
45040 LOT=0:RETURN
45050 LOCATE 17,10:PRINT "NAME OF THE READING FILE:"; :INPUT
NN$
45060 OPEN NN$ FOR INPUT AS #1
45070 INPUT #1,QUI:DIM QP(QUI,4),QU(QUI,4)
45080 FOR I=1 TO QUI:FOR J=1 TO 4:INPUT #1,QP(I,J):INPUT
#1,QU(I,J):NEXT J,I
45090 CLOSE
45100 LOT=1:RETURN
```

```
90 KEY OFF
100 REM           ****************************************
110 REM           *           PROGRAM   'PMSB5'          *
120 REM           *    WRITTEN BY DR.A./R.A. LATTERMANN   *
130 REM           *                1987                  *
140 REM           ****************************************
150 REM
160 CLS:BEEP
162 REM
163 REM UTILIZATION OF THE PROGRAM AND INPUT OF DATA
165 REM
170 GOSUB 2000
180 PRINT
190 PRINT TAB(10)"THE PROGRAM 'PMSB5' CALCULATES THE MEAN AREAL
PRECIPITATION"
200 PRINT TAB(10)"USING THE K R I G I N G - METHOD"
210 PRINT
220 PRINT TAB(10)"DATA OF INPUT ARE :"
230 PRINT
240 PRINT TAB(10)"1.    MAX. NUMBER OF TIME INTERVALS OF THE
CONSIDERED"
250 PRINT TAB(10)"      PRECIPITATION EVENT              NI=";
:INPUT NI
260 PRINT TAB(10)"2.    MAX. NUMBER OF GAGE-STATIONS  PI=";
:INPUT PI
265 PRINT TAB(10)"3.    MAX. NUMBER OF GRID-POINTS    GI=";
:INPUT GPI
270 PRINT TAB(10)"4.    MEASURED PRECIPITATION VALUES:"
280 A=NI:IF NI<PI THEN A=PI:IF A<GPI THEN A=GPI
290 DIM
P(A,A),PM(A),XS(A),YS(A),LDS(A,A),GAM(A+2,A+2),GAMBB(A+2),XG(GP
I),YG(GPI),LDG(A,GPI),GAMB(A,GPI),C(A+2),PHI(A+1,A+1),S(A+2),SI
G(A)
300 FOR I=1 TO PI:LOCATE 15,10:PRINT "STATION "; I:FOR J=1 TO
NI
310 LOCATE 16,10:PRINT "MEASURED PRECIPITATION (MM) OF INTERVAL
"; J; " =                   ";:LOCATE 16,55:INPUT P(I,J):NEXT J,I
311 PRINT:CLS:PRINT
312 LOCATE 8,10:PRINT "5. COORDINATES OF THE GAGE-STATIONS:"
314 FOR I=1 TO PI
316 LOCATE 10,10:PRINT "STATION"; I;": X-COORDINATE =";:LOCATE
10,40:INPUT XS(I):LOCATE 11,22:PRINT "Y-COORDINATE =";:LOCATE
11,40:INPUT YS(I):NEXT I
320 LOCATE 13,10:PRINT "6. COORDINATES OF THE GRID-POINTS:"
321 FOR I=1 TO GPI
322 LOCATE 15,10:PRINT "GRID-POINT";I;": X-COORDINATE
=";:LOCATE 15,40:INPUT XG(I):LOCATE 16,25:PRINT "Y-COORDINATE
=";:LOCATE 16,40:INPUT YG(I):NEXT I
327 REM
330 REM LINEAR DISTANCES
```

```
335 GOSUB 13000
337 ZINT=0
340 ZINT=ZINT+1
342 REM SEMI-VARIOGRAM/KRIGING SYSTEM
345 GOSUB 14000
350 REM
352 PM(ZINT)=0:SIG(ZINT)=0
354 FOR I=1 TO PI
356 PM(ZINT)=PM(ZINT)+P(I,ZINT)*PHI(I,0)
357 SIG(ZINT)=SIG(ZINT)+GAM(I,PI+2)*PHI(I,0)
358 NEXT I
359 SIG(ZINT)=SQR(SIG(ZINT)+PHI(PI+1,0)-GAMNN)
360 IF ZINT <= NI THEN 340
365 REM
370 REM
380 REM
390 CLS:BEEP
400 GOSUB 2000
401 REM
402 REM DISPLAY
403 REM
410 PRINT:PRINT
420 PRINT TAB(10)"MEAN AREAL PRECIPITATION (INTERVAL)
(MM):":PRINT
430 FOR I=1 TO NI
440 PRINT TAB(10)"PM("; I; ")    =    "; :PRINT USING "###.###";
PM(I); :PRINT"   SIGMA="; :PRINT USING "###.##"; SIG(I): NEXT I
450 PRINT
460 PRINT TAB(10)"PRINTING OF THE RESULT (Y/N) ???"
470 U$=INKEY$
480 IF U$="" THEN 470
490 IF U$="Y" OR U$="y" THEN GOSUB 3000
495 GOTO 12000
500 CLS
510 GOSUB 2000
520 PRINT:PRINT
530 PRINT TAB(10)"=========END OF THE PROGRAM========="
540 KEY ON:PLAY "O4T200MLL8GAGAL2C":END
2000 PRINT:PRINT TAB(10)"[ PROGRAM :   P  M  S  B  5  ]
"; DATE$:PRINT:RETURN
3000 REM
3001 REM PRINTER
3002 REM
3005 LPRINT:LPRINT
3010 LPRINT CHR$(14); "              P  M  S  B  5";
CHR$(20):LPRINT
3020 LPRINT CHR$(15); "
WRITTEN BY DR.A./R.A. LATTERMANN"; CHR$(18)
3030 LPRINT:LPRINT:LPRINT
3038 LPRINT TAB(10)"
"; DATE$:LPRINT
3040 LPRINT TAB(10)"THE PROGRAM 'PMSB5' CALCULATES THE MEAN
AREAL PRECIPITATION"
```

```
3050 LPRINT TAB(10)"USING THE K R I G I N G -METHOD"
3060 LPRINT:LPRINT
3070 LPRINT TAB(10)"DATA OF INPUT ARE :":LPRINT
3080 LPRINT TAB(10)"1.    MAX. NUMBER OF TIME INTERVALS OF THE
CONSIDERED"
3090 LPRINT TAB(10)"       PRECIPITATION EVENT                NI = ";
NI:LPRINT
3100 LPRINT TAB(10)"2.    MAX. NUMBER OF GAGE-STATIONS  PI = ";
PI:LPRINT
3105 LPRINT TAB(10)"3.    MAX. NUMBER OF GRID-POINTS    GI = ";
GPI:LPRINT
3110 LPRINT TAB(10)"4.    MEASURED PRECIPITATION VALUES:"
3120 FOR I=1 TO PI:LPRINT:LPRINT TAB(10)"STATION :"; I
3130 LPRINT TAB(10)"INTERVAL:"; TAB(45)"MEASURED PRECIPITATION
(MM):"
3140 FOR J=1 TO NI
3150 LPRINT TAB(13)J; TAB(50)P(I,J)
3160 NEXT J,I:LPRINT
3162 LPRINT TAB(10)"5.    COORDINATES OF THE GAGE-STATIONS
:"
3164 LPRINT:LPRINT TAB(10)"STATION:"; TAB(35)"COORDINATES:    X
Y   :"
3166 FOR I=1 TO PI
3167 LPRINT TAB(13)I; TAB(50)XS(I); TAB(55)YS(I)
3168 NEXT I
3170 LPRINT
3172 LPRINT TAB(10)"6.    COORDINATES OF THE GRID-POINTS
:"
3173 LPRINT:LPRINT TAB(10)"GRID-POINT:"; TAB(35)"COORDINATES:
X    Y    :"
3175 FOR I=1 TO GPI
3176 LPRINT TAB(13)I; TAB(50)XG(I); TAB(55)YG(I)
3177 NEXT I
3178 LPRINT:LPRINT
3180 LPRINT TAB(10)"RESULTS     :":LPRINT
3190 LPRINT:LPRINT TAB(10)"MEAN AREAL PRECIPITATION (INTERVAL)
:":LPRINT
3200 LPRINT TAB(10)"INTERVAL:"; TAB(25)"MEAN AREAL
PRECIPIT.(MM):"; TAB(52)"SIGMA:"
3210 FOR I=1 TO NI
3220 LPRINT TAB(13); I; TAB(30); PM(I); TAB(50); SIG(I)
3230 NEXT I
3240 RETURN
12000 REM
12002 REM END
12005 REM
12009 PRINT
12010 PRINT TAB(10)"==== (N)EW CALCULATION OR (E)ND ??? ===="
12020 U$=INKEY$
12030 IF U$="" THEN 12020
12040 IF U$="N" OR U$="n" THEN RUN
12050 GOTO 500
12060 END
```

```
13000 FOR I=1 TO PI
13010 FOR J=1 TO PI
13020 LDS(I,J)=SQR((XS(I)-XS(J))^2+(YS(I)-YS(J))^2)
13030 NEXT J
13040 NEXT I
13050 FOR I=1 TO PI
13060 FOR J=1 TO GPI
13070 LDG(I,J)=SQR((XS(I)-XG(J))^2+(YS(I)-YG(J))^2)
13080 NEXT J
13090 NEXT I
13100 REM NUMBER OF DISTANCES
13110 NDS=PI*PI/2
13120 NDG=PI*GPI/2
13130 REM NUMBER OF CLASSES
13140 NOC=5
13150 REM CLASS WIDTHS
13160 MAX=0
13170 FOR I=1 TO PI
13180 FOR J=1 TO PI
13190 IF MAX <= LDS(I,J) THEN MAX=LDS(I,J)
13200 NEXT J
13210 NEXT I
13220 FOR I=1 TO PI
13230 FOR J=1 TO GPI
13240 IF MAX <= LDG(I,J) THEN MAX=LDG(I,J)
13250 NEXT J
13260 NEXT I
13270 WOC=MAX/5
13280 RETURN
14000 REM ESTIMATION OF THE SEMI-VARIOGRAM
14010 FOR K=1 TO 5
14020 GAMMA(K)=0
14030 ZAEL=0
14040 FOR I=1 TO PI
14050 FOR J=1 TO PI
14060 IF LDS(I,J)<=(K-1)*WOC OR LDS(I,J)>K*WOC THEN 14090
14070 ZAEL=ZAEL+1
14080 GAMMA(K)=GAMMA(K)+(P(I,ZINT)-P(J,ZINT))^2
14090 NEXT J
14100 NEXT I
14105 IF ZAEL=0 THEN ZAEL=1
14110 GAMMA(K)=GAMMA(K)/(2*ZAEL)
14120 NEXT K
14130 REM LINEAR REGRESSION FOR SEMI-VARIOGRAMM POINTS
14140 EGAM=0
14150 FOR I=1 TO 5
14160 EGAM=EGAM+GAMMA(I)
14170 NEXT I
14180 EGAM=EGAM/5
14190 EGAMQ=0
14200 FOR I=1 TO 5
14210 EGAMQ=EGAMQ+GAMMA(I)^2
14220 NEXT I
```

```
14230 EGAMQ=EGAMQ/5
14240 EWOC=0
14250 FOR I=1 TO 5
14260 EWOC=EWOC+WOC*(I-.5)
14270 NEXT I
14280 EWOC=EWOC/5
14290 EWOCQ=0
14300 FOR I=1 TO 5
14310 EWOCQ=EWOCQ+(WOC*(I-.5))^2
14320 NEXT I
14330 EWOCQ=EWOCQ/5
14340 EGAWO=0
14350 FOR I=1 TO 5
14360 EGAWO=EGAWO+GAMMA(I)*WOC*(I-.5)
14365 NEXT I
14370 EGAWO=EGAWO/5
14380 C1=(EGAWO-EGAM*EWOC)/(EWOCQ-EWOC^2)
14390 C2=EGAM-C1*EWOC
14400 REM SEMI-VARIANCE BETWEEN THE GAGE STATIONS
14410 FOR I=1 TO PI
14420 FOR J=1 TO PI
14430 GAM(I,J)=C1*LDS(I,J)+C2
14440 NEXT J
14450 NEXT I
14460 REM SEMI-VARIANCE WITHIN THE BLOCK
14470 FOR I=1 TO PI
14480 FOR J=1 TO GPI
14490 GAMB(I,J)=C1*LDG(I,J)+C2
14500 NEXT J
14510 NEXT I
14520 FOR I=1 TO PI
14530 GAMBB(I)=0
14540 FOR J=1 TO GPI
14550 GAMBB(I)=GAMBB(I)+GAMB(I,J)
14560 NEXT J
14570 GAMBB(I)=GAMBB(I)/GPI
14580 NEXT I
14590 REM KRIGING EQUATIONS
14600 FOR I=1 TO PI
14610 GAM(I,PI+1)=1
14620 NEXT I
14630 FOR I=1 TO PI
14640 GAM(PI+1,I)=1
14650 NEXT I
14660 FOR I=1 TO PI+1:GAM(I,I)=0: NEXT I:F=0
14670 FOR I=1 TO PI+1:GAM(I,PI+2)=GAMBB(I):NEXT
I:GAM(PI+1,PI+2)=1
14680 G=1E-10
14690 FOR KK=1 TO PI+1
14700 REM pivot
14710 FOR I=1 TO PI+1
14720 IF GAM(I,0)>0 THEN 14750
14730 S=0:FOR L=1 TO PI+1
```

```
14740 S=S+ABS(GAM(I,L))^2:NEXT L:S(I)=SQR(S)
14750 NEXT I
14760 P=0:FOR K=1 TO PI+1:FOR I=1 TO PI+1
14770 IF GAM(0,K)>0 THEN 14830
14780 IF GAM(I,K)=0 THEN 14820
14790 S=ABS(GAM(I,K))/S(I)
14800 IF S<=P OR GAM(I,0)>0 OR S(I)=0 THEN 14820
14810 P=S:IP=I:KP=K
14820 NEXT I
14830 NEXT K
14840 IF P=0 THEN 14960
14850 REM
14860 I=IP:K=KP
14870 GAM(0,K)=I:GAM(I,0)=K:F=F+1
14880 FOR L=1 TO PI+2:C(L)=GAM(I,L):NEXT L
14890 FOR J=1 TO PI+1
14900 Y=GAM(J,K)/C(K)
14910 FOR L=1 TO PI+2
14920 GAM(J,L)=GAM(J,L)-C(L)*Y
14930 IF ABS(GAM(J,L))<G THEN GAM(J,L)=0
14940 NEXT L:NEXT J
14950 FOR L=1 TO PI+2:GAM(I,L)=C(L)/C(K):NEXT L
14960 NEXT KK
14970 REM SOLUTION
14980 R=PI+1-F
14990 FOR K=1 TO PI+1
15000 I=GAM(0,K)
15010 IF I<1 THEN V=V+1:PHI(K,V)=1:GOTO 15060
15015 PHI(K,0)=GAM(I,PI+2)
15020 FOR L=1 TO PI+1
15030 IF GAM(0,1)>0 THEN 15050
15040 W=W+1:PHI(K,W)=-GAM(I,L)
15050 NEXT L:W=0
15060 NEXT K
15070 REM
15080 REM TEST OF THE SOLUTION
15090 FOR I=1 TO PI+1
15100 IF (GAM(I,0)=0 AND GAM(I,PI+2)<>0) THEN PRINT"NO UNIQUE
SOLUTION":END
15110 NEXT I
15115 GAMNN=0
15120 FOR I=1 TO PI
15130 GAMNN=GAMNN+GAM(I,PI+2)
15140 NEXT I
15150 GAMNN=GAMNN/PI
15160 RETURN
```

```
90 KEY OFF
100 REM            *****************************************
110 REM            *          PROGRAM   'PREF1'            *
120 REM            *    WRITTEN BY DR.A./R.A. LATTERMANN   *
130 REM            *                1987                   *
140 REM            *****************************************
150 REM
160 CLS:BEEP
161 REM
162 REM UTILIZATION OF THE PROGRAM AND DATA INPUT
163 REM
170 GOSUB 2000
180 PRINT:PRINT
190 PRINT TAB(10)"THE PROGRAM 'PREF1' CALCULATES VALUES OF THE
EFFECTIVE RAIN"
200 PRINT TAB(10)"BY USE OF THE Φ-I N D E X  METHOD"
210 PRINT
220 PRINT TAB(10)"dATA OF INPUT ARE:"
240 PRINT TAB(10)"1.      CATCHMENT AREA                 (QKM) = ";
:INPUT S
260 PRINT TAB(10)"2.      VOLUME OF THE SURFACE RUNOFF(CBM) = ";
:INPUT Q
265 QMM=Q/(1000*S)
270 PRINT TAB(10)"3.      MAX. NUMBER OF TIME INTERVALS     = ";
:INPUT N
280 DIM P(N),PE(N)
300 FOR I=1 TO N:LOCATE 13,17:PRINT "INTERVAL "; I
310 LOCATE 14,17:PRINT "MEAN AREAL PRECIPITATION (MM)
";:LOCATE 14,55   :INPUT P(I):NEXT I
312 FOR I=1 TO N:SP=SP+P(I):NEXT I
315 GOSUB 8000
390 CLS:BEEP
391 REM
392 REM FOR DISPLAY ..
393 REM
400 GOSUB 2000
410 PRINT:PRINT
430 PRINT TAB(10)"MEAN AREAL PRECIPIT.(MM)"; TAB(35)"EFFECTIVE
RAIN VALUE (MM)"
440 FOR I=1 TO N
445 PRINT TAB(16); :PRINT USING "###.##"; P(I); :PRINT TAB(45);
:PRINT USING "###.##"; PE(I):NEXT I
446 PRINT
TAB(10)"==================================================="
447 PRINT TAB(10)"TOTAL:"; TAB(16); :PRINT USING "###.##"; SP;
:PRINT TAB(45); :PRINT USING "###.##"; SE
450 PRINT
460 PRINT TAB(10)"==== PRINTING OF THE RESULT (Y/N) ???"
470 U$=INKEY$
480 IF U$="" THEN 470
```

```
490 IF U$="Y" OR U$="y" THEN GOSUB 3000
495 GOTO 12000
500 CLS
510 GOSUB 2000
520 PRINT:PRINT
530 PRINT TAB(10)"========== END OF THE PROGRAM ========="
540 KEY ON:PLAY "O4T200MLL8GAGAL2C":END
2000 PRINT TAB(10)"[ PROGRAM :     P   R   E   F   1   ]
"; DATE$:PRINT:RETURN
3000 REM
3001 REM PRINTER ..
3002 REM
3009 LPRINT:LPRINT
3010 LPRINT CHR$(14); "              P   R   E   F   1";
CHR$(20):LPRINT
3020 LPRINT CHR$(15); "
WRITTEN BY DR.A./R.A. LATTERMANN"; CHR$(18)
3030 LPRINT:LPRINT:LPRINT
3038 LPRINT TAB(10)"
"; DATE$:LPRINT
3040 LPRINT TAB(10)"THE PROGRAM 'PREF1' CALCULATES VALUES OF
THE EFFECTIVE RAIN"
3050 LPRINT TAB(10)"BY USE OF THE Φ-INDEX METHOD"
3060 LPRINT:LPRINT
3070 LPRINT TAB(10)"DATA OF INPUT ARE:":LPRINT
3080 LPRINT TAB(10)"1.    CATCHMENT AREA                    (QKM) :";
S:LPRINT
3100 LPRINT TAB(10)"2.    VOLUME OF THE SURFACE RUNOFF (CBM) :";
Q
3110 LPRINT:LPRINT TAB(10)"3.    GUESS FOR   Φ
(MM) :"; FI
3115 LPRINT:LPRINT TAB(10)"4.    MEAN AREAL PRECIPITATION
(MM) :"
3120 LPRINT:LPRINT TAB(15)"INTERVAL"; TAB(45)"MEAN AREAL
PRECIPIT.(MM)"
3140 FOR I=1 TO N
3150 LPRINT TAB(18)I; TAB(55)P(I)
3160 NEXT I
3170 LPRINT:LPRINT:LPRINT
3180 LPRINT TAB(10)"RESULTS:":LPRINT
3200 LPRINT TAB(10)"INTERVAL:"; TAB(25)"MEAN AREAL
PRECIPIT.(MM):" ; TAB(47)"EFFECTIVE RAIN VALUE (MM):"
3210 FOR I=1 TO N
3220 LPRINT TAB(13)I; TAB(32)P(I); TAB(55)PE(I)
3230 NEXT I
3232 LPRINT
TAB(10)"=======================================================
====="
3233 LPRINT TAB(10)"TOTAL:"; TAB(32)SP; TAB(55)SE
3240 RETURN
8000 REM
8001 REM CALCULATION ..
8003 REM
```

```
8010 LOCATE 19,17:PRINT "GUESS FOR Φ      (MM) = "; :INPUT FI
8030 FOR U=INT(FI/2) TO INT(3*FI/2) STEP .05
8040 FOR I=1 TO N
8050 PE(I)=P(I)-U
8060 IF PE(I)<0 THEN PE(I)=0
8070 SE=SE+PE(I)
8080 NEXT I
8090 IF (SE-QMM)<0 THEN RETURN
8100 SE=0
8110 NEXT U
8125 PLAY "O4L32T100MSACACACACAC"
8130 PRINT TAB(17)"DESIRED PRECISION CANNOT BE ACHIEVED
!":PRINT TAB(17)"VARIATION OF I = "; FII; " TO "; U-1
8140 PRINT TAB(17)"NEW TRIAL   (Y/N) ???"
8150 U$=INKEY$
8160 IF U$="" THEN 8150
8170 IF U$="Y" OR U$="y" THEN LOCATE 20,17:PRINT
SPACE$(60):LOCATE 21,17:PRINT SPACE$(60):LOCATE 22,17:PRINT
SPACE$(60):GOTO 8000
8180 CLS:GOSUB 2000:PLAY "O4T200L8MLGAGAL2C":PRINT:PRINT:PRINT
TAB(10)"====>   E   N   D   <====":KEY ON:END
12000 REM
12001 REM END ..
12002 REM
12009 PRINT
12010 PRINT TAB(10)"==== (N)EW CALCULATION OR (E)ND ??? ===="
12020 U$=INKEY$
12030 IF U$="" THEN 12020
12040 IF U$="N" OR U$="n" THEN RUN
12050 GOTO 500
12060 END
```

```
90 KEY OFF
100 REM      ******************************************
110 REM      *           PROGRAM  'PREF2'             *
120 REM      *    WRITTEN BY DR.A./R.A. LATTERMANN    *
130 REM      *                1987                    *
140 REM      ******************************************
150 REM
160 CLS:BEEP
161 REM
162 REM UTILIZATION OF THE PROGRAM AND DATA INPUT
163 REM
170 GOSUB 2000
180 PRINT:PRINT
190 PRINT TAB(10)"THE PROGRAM 'PREF2' CALCULATES VALUES OF THE
EFFECTIVE RAIN"
200 PRINT TAB(10)"BY USE OF THE METHOD OF  P E R C E N T A G E
LOSS RATES "
210 PRINT
220 PRINT TAB(10)"DATA OF INPUT ARE:"
240 PRINT TAB(10)"1.      CATCHMENT AREA              (QKM) =
"; :INPUT S
260 PRINT TAB(10)"2.      VOLUME OF THE SURFACE RUNOFF (CBM) =
"; :INPUT Q
265 QMM=Q/(1000*S)
270 PRINT TAB(10)"3.      MAX. NUMBER OF TIME INTERVALS      =
"; :INPUT N
280 DIM P(N),PE(N)
300 FOR I=1 TO N:LOCATE 13,17:PRINT "INTERVAL:"; I
310 LOCATE 14,17:PRINT "MEAN AREAL PRECIPITATION (MM)
";:LOCATE 14,55  :INPUT P(I):NEXT I
312 FOR I=1 TO N:SP=SP+P(I):NEXT I
315 GOSUB 8000
390 CLS:BEEP
391 REM
392 REM FOR DISPLAY ..
393 REM
400 GOSUB 2000
410 PRINT:PRINT
420 PRINT TAB(10)"EFFECTIVE RAIN VALUE (INTERVAL) (MM):":PRINT
430 PRINT TAB(10)"MEAN AREAL PRECIPIT."; TAB(35)"EFFECTIVE RAIN
VALUE"
440 FOR I=1 TO N
445 PRINT TAB(16); :PRINT USING "###.##"; P(I); :PRINT TAB(45);
:PRINT USING "###.##"; PE(I):NEXT I
446 PRINT
TAB(10)"==================================================="
447 PRINT TAB(10)"TOTAL:"; TAB(16); :PRINT USING "###.##"; SP;
:PRINT TAB(45); :PRINT USING "###.##"; SE
450 PRINT
460 PRINT TAB(10)"==== PRINTING OF THE RESULT (Y/N) ???"
```

```
470 U$=INKEY$
480 IF U$="" THEN 470
490 IF U$="Y" OR U$="y" THEN GOSUB 3000
495 GOTO 12000
500 CLS
510 GOSUB 2000
520 PRINT:PRINT
530 PRINT TAB(10)"========== END OF THE PROGRAM ========"
540 KEY ON:PLAY "O4T200L8MLGAGAL2C":END
2000 PRINT TAB(10)"[ PROGRAM :    P   R   E   F   2   ]
"; DATE$:PRINT:RETURN
3000 REM
3001 REM PRINTER  ..
3002 REM
3009 LPRINT:LPRINT
3010 LPRINT CHR$(14); "                P   R   E   F   2";
CHR$(20):LPRINT
3020 LPRINT CHR$(15); "
WRITTEN BY DR.A./R.A. LATTERMANN"; CHR$(18)
3030 LPRINT:LPRINT:LPRINT
3038 LPRINT TAB(10)"
"; DATE$:LPRINT
3040 LPRINT TAB(10)"THE PROGRAM 'PREF2' CALCULATES VALUES OF
THE EFFECTIVE RAIN"
3050 LPRINT TAB(10)"BY USE OF THE METHOD OF PERCENTAGE LOSS
RATES "
3060 LPRINT:LPRINT
3070 LPRINT TAB(10)"DATA OF INPUT ARE:":LPRINT
3080 LPRINT TAB(10)"1.    CATCHMENT AREA                  (QKM) :";
S:LPRINT
3100 LPRINT TAB(10)"2.    VOLUME OF THE SURFACE RUNOFF (CBM) :";
Q
3115 LPRINT:LPRINT TAB(10)"3.    MEAN AREAL PRECIPITATION
(MM) :"
3120 LPRINT:LPRINT TAB(15)"INTERVAL"; TAB(45)"MEAN AREAL
PRECIPIT.(MM)"
3140 FOR I=1 TO N
3150 LPRINT TAB(18)I; TAB(55)P(I)
3160 NEXT I
3170 LPRINT:LPRINT:LPRINT
3180 LPRINT TAB(10)"RESULTS:":LPRINT
3190 LPRINT:LPRINT TAB(10)"EFFECTIVE RAIN VALUE (INTERVAL)
:":LPRINT
3200 LPRINT TAB(10)"INTERVAL:"; TAB(25)"MEAN AREAL PRECIPIT." ;
TAB(47)"EFFECTIVE RAIN VALUE:"
3210 FOR I=1 TO N
3220 LPRINT TAB(13)I; TAB(32)P(I); TAB(55)PE(I)
3230 NEXT I
3232 LPRINT
TAB(10)"===============================================
===="
3233 LPRINT TAB(10)"TOTAL:"; TAB(32)SP; TAB(55)SE
3240 RETURN
```

```
8000 REM
8001 REM CALCULATION ..
8002 REM
8010 K=1-QMM/SP
8020 SE=0:FOR I=1 TO N
8030 PE(I)=P(I)-P(I)*K
8040 SE=SE+PE(I)
8050 NEXT I
8060 RETURN
12000 REM
12001 REM END ..
12002 REM
12009 PRINT
12010 PRINT TAB(10)"==== (N)EW CALCULATION OR (E)ND ??? ===="
12020 U$=INKEY$
12030 IF U$="" THEN 12020
12040 IF U$="N" OR U$="n" THEN RUN
12050 GOTO 500
12060 END
```

```
90 KEY OFF
100 REM           ******************************************
110 REM           *          PROGRAM    'PREF3'            *
120 REM           *    WRITTEN BY DR.A./R.A. LATTERMANN    *
130 REM           *                 1987                   *
140 REM           ******************************************
150 REM
160 CLS:BEEP
161 REM
162 REM UTILIZATION OF THE PROGRAM AND DATA INPUT
163 REM
170 GOSUB 2000
180 PRINT:PRINT
190 PRINT TAB(10)"THE PROGRAM 'PREF3' CALCULATES VALUES OF THE
EFFECTIVE RAIN"
200 PRINT TAB(10)"BY USE OF THE H O R T O N  METHOD"
210 PRINT
220 PRINT TAB(10)"DATA OF INPUT ARE:"
240 PRINT TAB(10)"1.      CATCHMENT AREA                 (QM) =
"; :INPUT S
260 PRINT TAB(10)"2.      VOLUME OF THE SURFACE RUNOFF (CBM) =
"; :INPUT Q
265 QMM=Q*1000/S
270 PRINT TAB(10)"3.      MAX. NUMBER OF TIME INTERVALS       =
"; :INPUT N
280 DIM P(N),PE(N)
300 FOR I=1 TO N:LOCATE 13,17:PRINT "INTERVAL "; I
310 LOCATE 14,17:PRINT "MEAN AREAL PRECIPITATION (MM)     ";:
LOCATE 14,55 :INPUT P(I):NEXT I
312 FOR I=1 TO N:SP=SP+P(I):NEXT I
315 GOSUB 8000
390 CLS:BEEP
391 REM
392 REM FOR DISPLAY ..
393 REM
400 GOSUB 2000
410 PRINT:PRINT
420 PRINT TAB(10)"EFFECTIVE RAIN VALUE (INTERVAL) (MM):":PRINT
430 PRINT TAB(10)"MEAN AREAL PRECIPIT."; TAB(35)"EFFECTIVE RAIN
VALUE"
440 FOR I=1 TO N
445 PRINT TAB(16); :PRINT USING "###.##"; P(I); :PRINT TAB(45);
:PRINT USING "###.##"; PE(I):NEXT I
446 PRINT
TAB(10)"================================================="
447 PRINT TAB(10)"TOTAL:"; TAB(16); :PRINT USING "###.##"; SP;
:PRINT TAB(45); :PRINT USING "###.##"; SE
450 PRINT
460 PRINT TAB(10)"==== PRINTING OF THE RESULT (Y/N) ???"
470 U$=INKEY$
```

```
480 IF U$="" THEN 470
490 IF U$="Y" OR U$="y" THEN GOSUB 3000
495 GOTO 12000
500 CLS
510 GOSUB 2000
520 PRINT:PRINT
530 PRINT TAB(10)"========== END OF THE PROGRAM =========="
540 KEY ON:PLAY "O4T200MLL8GAGAL2C":END
2000 PRINT TAB(10)"[ PROGRAM :    P  R  E  F  3   ]
"; DATE$:PRINT:RETURN
3000 REM
3001 REM PRINTER ..
3002 REM
3009 LPRINT:LPRINT
3010 LPRINT CHR$(14); "              P  R  E  F  3";
CHR$(20):LPRINT
3020 LPRINT CHR$(15); "
WRITTEN BY DR.A./R.A. LATTERMANN"; CHR$(18)
3030 LPRINT:LPRINT:LPRINT
3038 LPRINT TAB(10)"
"; DATE$:LPRINT
3040 LPRINT TAB(10)"THE PROGRAM 'PREF3' CALCULATES VALUES OF
THE EFFECTIVE RAIN"
3050 LPRINT TAB(10)"BY USE OF THE HORTON METHOD"
3060 LPRINT:LPRINT
3070 LPRINT TAB(10)"DATA OF INPUT ARE:":LPRINT
3080 LPRINT TAB(10)"1.   CATCHMENT AREA                      (QKM)
:"; S:LPRINT
3100 LPRINT TAB(10)"2.   VOLUME OF OF THE SURFACE RUNOFF (CBM)
:"; Q
3110 LPRINT:LPRINT TAB(10)"3.   A.) INITIAL INFILTRATION
CAPACITY      (KO)"; K
3111 LPRINT TAB(10)"        B.) STATIONARY INFILTRATION CAPACITY
(KC)"; KC
3112 LPRINT TAB(10)"        C.) COEFFICIENT ALFA
"; ALFA
3115 LPRINT:LPRINT TAB(10)"4.   MEAN AREAL PRECIPITATION   :"
3120 LPRINT:LPRINT TAB(15)"INTERVAL"; TAB(45)"MEAN AREAL
PRECIPIT.(MM)"
3140 FOR I=1 TO N
3150 LPRINT TAB(18)I; TAB(55)P(I)
3160 NEXT I
3170 LPRINT:LPRINT:LPRINT
3180 LPRINT TAB(10)"RESULTS:":LPRINT
3190 LPRINT:LPRINT TAB(10)"EFFECTIVE RAIN VALUE (INTERVAL) [MM]
:":LPRINT
3200 LPRINT TAB(10)"INTERVAL:"; TAB(25)"MEAN AREAL PRECIPIT.:"
; TAB(47)"EFFECTIVE RAIN VALUE:"
3210 FOR I=1 TO N
3220 LPRINT TAB(13)I; TAB(32)P(I); TAB(55)PE(I)
3230 NEXT I
```

```
3232 LPRINT
TAB(10)"=================================================================
====="
3233 LPRINT TAB(10)"TOTAL:"; TAB(32)SP; TAB(55)SE
3240 RETURN
8000 REM
8001 REM CALCULATION ..
8002 REM
8010 LOCATE 19,17:PRINT "INITIAL INFILTRATION CAPACITY      (K)";
:INPUT K
8011 LOCATE 19,17:PRINT SPACE$(50)
8012 LOCATE 19,17:PRINT "STATIONARY INFILTRATION CAPACITY
(KC)"; :INPUT KC
8013 LOCATE 19,17:PRINT SPACE$(50)
8014 LOCATE 19,17:PRINT "COEFFICIENT ALFA
"; :INPUT ALFA
8040 FOR I=1 TO N
8050 B=-ALFA*I
8060 PE(I)=P(I)-(KC+(K-KC)*EXP(B)):IF PE(I)<0 THEN PE(I)=0
8070 SE=SE+PE(I)
8080 NEXT I
8090 RETURN
12000 REM
12001 REM END ..
12002 REM
12009 PRINT
12010 PRINT TAB(10)"==== (N)EW CALCULATION OR (E)ND ??? ===="
12020 U$=INKEY$
12030 IF U$="" THEN 12020
12040 IF U$="N" OR U$="n" THEN RUN
12050 GOTO 500
12060 END
```

```
90 KEY OFF
100 REM         ****************************************
110 REM         *            PROGRAM  'VAZAO1'          *
120 REM         *     WRITTEN BY DR.A./R.A. LATTERMANN  *
130 REM         *                 1987                  *
140 REM         ****************************************
150 REM
160 CLS:BEEP
161 REM
162 REM UTILIZATION OF THE PROGRAM AND DATA INPUT
163 REM
170 GOSUB 2000
180 PRINT:PRINT
190 PRINT TAB(10)"THE PROGRAM 'VAZAO1' CALCULATES RUNOFF
VALUES"
200 PRINT TAB(10)"BY USE OF THE M A T R I X METHOD"
210 PRINT
220 PRINT TAB(10)"DATA OF INPUT ARE:":PRINT
240 PRINT TAB(10)"1.      TIME DISTRIBUTION OF EFFECTIVE RAIN"
250 PRINT TAB(10)"     A.) MAX. NUMBER OF INTERVALS MI":PRINT
TAB(10)"      B.) EFFECTIVE RAIN VALUE (MM) / INTERVAL":PRINT
TAB(10)"          N0,N1,N2,N3,......"
260 PRINT:PRINT TAB(10)"2.      SYSTEM RESPONSE FUNCTION"
265 PRINT TAB(10)"     A.) MAX. NUMBER OF TIME INTERVALS
SI":PRINT TAB(10)"      B.) VALUES OF SYSTEM RESPONSE (MM) /
INTERVAL":PRINT TAB(10)"          H0,H1,H2,H3,......"
270 PRINT:LOCATE 20,10:PRINT "     MAX. NUMBER OF TIME
INTERVALS MI = "; :INPUT MI:M=MI-1:DIM NI(MI)
300 FOR I=0 TO M:LOCATE 21,17:PRINT "EFFECTIVE RAIN VALUE N"; I
310 LOCATE 22,10:PRINT "
";:LOCATE 22,55:INPUT NI(I):NEXT I:LOCATE 20,10:PRINT
SPACE$(45)
312 LOCATE 21,10:PRINT SPACE$(50):LOCATE 22,10:PRINT
SPACE$(50):LOCATE 20,10:PRINT "        MAX. NUMBER OF TIME
INTERVALS SI = "; :INPUT SI:S=SI-1:DIM HI(S+M)
313 FOR I=0 TO S:LOCATE 21,17:PRINT "SYSTEM RESPONSE FUNCTION
(UH) H"; I:L=M+S
314 LOCATE 22,10:PRINT "
";:LOCATE 14,55 :INPUT HI(I):NEXT I:PLAY "O2MLT250L8CEGL4O3C"
315 CLS:LOCATE 13,25:PRINT "W A I T   ! ! !":GOSUB 8000
390 CLS:BEEP
391 REM
392 REM DISPLAY ..
393 REM
400 GOSUB 2000
410 PRINT:PRINT
420 PRINT TAB(10)"RUNOFF CALCULATED        :":PRINT
430 PRINT TAB(10)"INTERVAL (I)"; TAB(35)"RUNOFF (Q(I))"
440 FOR I=0 TO L
```

```
445 PRINT TAB(16)I; TAB(40); :PRINT USING "###.###"; Q(I):NEXT
I
450 PRINT
460 PRINT TAB(10)"==== PRINTING OF THE RESULT (Y/N) ??? ===="
470 U$=INKEY$
480 IF U$="" THEN 470
490 IF U$="Y" OR U$="y" THEN GOSUB 3000
495 GOSUB 20000:GOTO 12000
500 CLS
510 GOSUB 2000
520 PRINT:PRINT
530 PRINT TAB(10)"========== END OF THE PROGRAM ========"
540 PLAY "O4T200L8GAGAL2C":KEY ON:END
2000 PRINT TAB(10)"[ PROGRAM :    V  A  Z  A  O  1  ]
"; DATE$:PRINT:RETURN
3000 REM
3001 REM PRINTER ..
3002 REM
3009 LPRINT:LPRINT
3010 LPRINT CHR$(14); "            V  A  Z  A  O  1";
CHR$(20):LPRINT
3020 LPRINT CHR$(15); "
WRITTEN BY DR.A./R.A. LATTERMANN"; CHR$(18)
3030 LPRINT:LPRINT:LPRINT
3038 LPRINT TAB(10)"
"; DATE$:LPRINT
3040 LPRINT TAB(10)"THE PROGRAM 'VAZAO1' CALCULATES RUNOFF
VALUES "
3050 LPRINT TAB(10)"BY USE OF THE MATRIX METHOD"
3060 LPRINT:LPRINT
3070 LPRINT TAB(10)"DATA OF INPUT ARE:":LPRINT
3080 LPRINT TAB(10)"1.    TIME DISTRIBUTION OF EFFECTIVE RAIN"
3090 LPRINT TAB(10)"     A.) MAX. NUMBER OF TIME INTERVALS
MI":LPRINT TAB(10)"     B.) EFFECTIVE RAIN VALUE (MM) /
INTERVAL":LPRINT TAB(10)"          N1,N2,N3, ..."
3100 LPRINT:LPRINT TAB(10)"2.    SYSTEM RESPONSE FUNCTION "
3105 LPRINT TAB(10)"     A.) MAX. NUMBER OF TIME INTERVALS
SI":LPRINT TAB(10)"     B.) SYSTEM RESPONSE VALUE   (MM) /
INTERVAL":LPRINT TAB(10)"          H1,H2,H3, ...":LPRINT
3110 LPRINT TAB(10)"MAX. NUMBER OF TIME INTERVALS MI = ";
MI:LPRINT
3120 LPRINT TAB(15)"INTERVAL"; TAB(45)"MEAN AREAL
PRECIPIT.(MM)"
3140 FOR I=0 TO M
3150 LPRINT TAB(18)I; TAB(55)NI(I)
3160 NEXT I
3162 LPRINT:LPRINT TAB(10)"MAX. NUMBER OF TIME INTERVALS SI =
"; SI:LPRINT
3164 LPRINT TAB(15)"INTERVAL"; TAB(45)"SYSTEM RESPONSE VALUES
(MM)"
3166 FOR I=0 TO S
3168 LPRINT TAB(18)I; TAB(58)HI(I)
3169 NEXT I
```

```
3170 LPRINT:LPRINT:LPRINT
3180 LPRINT TAB(10)"RESULTS :":LPRINT
3190 LPRINT:LPRINT TAB(10)"RUNOFF VALUE (MM) / INTERVAL
[Q(I)]:":LPRINT
3200 LPRINT TAB(10)"INTERVAL:"; TAB(45)"Q(I)"
3210 FOR I=0 TO L
3220 LPRINT TAB(13)I; TAB(45)Q(I)
3230 NEXT I
3232 LPRINT:LPRINT TAB(10)"E   N   D ..."
3240 RETURN
8000 REM
8001 REM CALCULATION ..
8002 REM
8010 DIM C(SI+2*MI,MI),Q(SI+MI)
8020 X=0
8030 FOR I=0 TO M
8040 FOR J=0 TO L-2
8050 C(X+J,I)=HI(J)
8060 NEXT J
8070 X=X+1 .
8080 NEXT I
8090 FOR I=0 TO L
8100 Q(I)=0
8110 FOR J=0 TO MI
8120 Q(I)=Q(I)+C(I,J)*NI(J)
8125 NEXT J,I
8130 RETURN
12000 REM
12001 REM END ..
12002 REM
12009 PRINT
12010 PRINT TAB(10)"==== (N)EW CALCULATION OR (E)ND ??? ===="
12020 U$=INKEY$
12030 IF U$="" THEN 12020
12040 IF U$="N" OR U$="n" THEN RUN
12050 GOTO 500
20000 END
```

```
90 KEY OFF
100 REM             *****************************************
110 REM             *             PROGRAM  'VAZAO2'          *
120 REM             *     WRITTEN BY DR.A./R.A. LATTERMANN   *
130 REM             *                 1987                   *
140 REM             *****************************************
150 REM
160 CLS:BEEP
161 REM
162 REM UTILIZATION OF THE PROGRAM AND DATA INPUT
163 REM
170 GOSUB 2000
180 PRINT:PRINT
190 PRINT TAB(10)"THE PROGRAM 'VAZAO2' CALCULATES RUNOFF
VALUES"
200 PRINT TAB(10)"BY USE OF THE METHOD OF Z-T R A N S F O R M A
T I O N"
210 PRINT
220 PRINT TAB(10)"DATA OF INPUT ARE:":PRINT
240 PRINT TAB(10)"1.      TIME DISTRIBUTION OF EFFECTIVE RAIN "
250 PRINT TAB(10)"       A.) MAX. NUMBER OF TIME INTERVALS
MI":PRINT TAB(10)"       B.) EFFECTIVE RAIN VALUE (MM) /
INTERVAL":PRINT TAB(10)"          N0,N1,N2,N3,......"
260 PRINT:PRINT TAB(10)"2.      SYSTEM RESPONSE FUNCTION "
265 PRINT TAB(10)"       A.) MAX. NUMBER OF TIME INTERVALS
SI":PRINT TAB(10)"       B.) VALUES OF SYSTEM RESPONSE (MM) /
INTERVAL":PRINT TAB(10)"          H0,H1,H2,H3,......"
270 PRINT:LOCATE 20,10:PRINT "      MAX. NUMBER OF TIME
INTERVALS MI = "; :INPUT MI:M=MI-1:DIM NI(M)
300 FOR I=0 TO M:LOCATE 21,17:PRINT "EFFECTIVE RAIN VALUE  N";
I
310 LOCATE 22,10:PRINT "
";:LOCATE 22,55  :INPUT NI(I):NEXT I:LOCATE 20,10:PRINT
SPACE$(45)
312 LOCATE 21,17:PRINT SPACE$(50):LOCATE 22,10:PRINT
SPACE$(50):LOCATE 20,10:PRINT "       MAX. NUMBER OF TIME
INTERVALS SI = "; :INPUT SI:S=SI-1:DIM HI(S)
313 FOR I=0 TO S:LOCATE 21,17:PRINT "SYSTEM RESPONSE FUNCTION
(UH) H"; I
314 LOCATE 22,10:PRINT "
";:LOCATE 22,55  :INPUT HI(I):NEXT I:PLAY "O2MLT250L8CEGL4O3C"
315 CLS:LOCATE 13,25:PRINT "W A I T  ! ! !":GOSUB 8000
390 CLS:BEEP
391 REM
392 REM DISPLAY ..
393 REM
400 GOSUB 2000
410 PRINT:PRINT
420 PRINT TAB(10)"RUNOFF CALCULATED         :":PRINT
430 PRINT TAB(10)"INTERVAL (I)"; TAB(35)"RUNOFF (Q(I))"
```

```
440 FOR I=0 TO L
445 PRINT TAB(16)I; TAB(40); :PRINT USING "###.###"; Q(I):NEXT
I
450 PRINT
460 PRINT TAB(10)"==== PRINTING OF THE RESULT (Y/N) ??? ===="
470 U$=INKEY$
480 IF U$="" THEN 470
490 IF U$="Y" OR U$="y" THEN GOSUB 3000
495 GOSUB 20000:GOTO 12000
500 CLS
510 GOSUB 2000
520 PRINT:PRINT
530 PRINT TAB(10)"========== END OF THE PROGRAM =========="
540 PLAY "O4T200L8GAGAL2C":KEY ON:END
2000 PRINT TAB(10)"[ PROGRAM :    V  A  Z  A  O  2  ]
"; DATE$:PRINT:RETURN
3000 REM
3001 REM PRINTER ..
3002 REM
3009 LPRINT:LPRINT
3010 LPRINT CHR$(14); "                V  A  Z  A  O  2";
CHR$(20):LPRINT
3020 LPRINT CHR$(15); "
WRITTEN BY DR.A./R.A. LATTERMANN"; CHR$(18)
3030 LPRINT:LPRINT:LPRINT
3038 LPRINT TAB(10)"
"; DATE$:LPRINT
3040 LPRINT TAB(10)"THE PROGRAM 'VAZAO2' CALCULATES RUNOFF
VALUES "
3050 LPRINT TAB(10)"BY THE USE OF THE METHOD OF Z-
TRANSFORMATION"
3060 LPRINT:LPRINT
3070 LPRINT TAB(10)"DATA OF INPUT ARE:":LPRINT
3080 LPRINT TAB(10)"1.    TIME DISTRIBUTION OF EFFECTIVE RAIN"
3090 LPRINT TAB(10)"    A.) MAX. NUMBER OF TIME INTERVALS
MI":LPRINT TAB(10)"    B.) EFFECTIVE RAIN VALUE (MM) /
INTERVAL":LPRINT TAB(10)"        N0,N1,N2,N3, ..."
3100 LPRINT:LPRINT TAB(10)"2.    SYSTEM RESPONSE FUNCTION "
3105 LPRINT TAB(10)"    A.) MAX. NUMBER OF TIME INTERVALS
SI":LPRINT TAB(10)"    B.) SYSTEM RESPONSE VALUE (MM) /
INTERVAL":LPRINT TAB(10)"        H0,H1,H2,H3, ...":LPRINT
3110 LPRINT TAB(10)"MAX. NUMBER OF TIME INTERVALS MI = ";
MI:LPRINT
3120 LPRINT TAB(15)"INTERVAL"; TAB(45)"EFFECTIVE RAIN VALUE
(MM)"
3140 FOR I=0 TO M
3150 LPRINT TAB(18)I; TAB(55)NI(I)
3160 NEXT I
3162 LPRINT:LPRINT TAB(10)"MAX. NUMBER OF TIME INTERVALS SI =
"; SI:LPRINT
3164 LPRINT TAB(15)"INTERVAL"; TAB(45)"SYSTEM RESPONSE VALUES
(MM)"
3166 FOR I=0 TO S
```

```
3168 LPRINT TAB(18)I; TAB(58)HI(I)
3169 NEXT I
3170 LPRINT:LPRINT:LPRINT
3180 LPRINT TAB(10)"RESULTS:":LPRINT
3190 LPRINT:LPRINT TAB(10)"RUNOFF VALUE (MM) / INTERVAL
[Q(I)]:":LPRINT
3200 LPRINT TAB(10)"INTERVAL:"; TAB(45)"Q(I)"
3210 FOR I=0 TO L
3220 LPRINT TAB(13)I; TAB(45)Q(I)
3230 NEXT I
3232 LPRINT:LPRINT TAB(10)"E   N   D   ..."
3240 RETURN
8000 REM
8001 REM CALCULATION ..
8002 REM
8010 DIM C(M,S)
8030 FOR I=0 TO M
8040 FOR J=0 TO S
8050 C(I,J)=HI(J)*NI(I)
8060 NEXT J
8080 NEXT I:N=0:L=M+S:DIM Q(L):FOR K=0 TO L:Q(K)=0
8090 FOR I=0 TO M
8110 FOR J=0 TO S:IF I+J=K THEN GOTO 8120
8115 GOTO 8125
8120 Q(K)=Q(K)+C(I,J)
8125 NEXT J,I,K
8130 RETURN
12000 REM
12001 REM END ..
12002 REM
12009 PRINT
12010 PRINT TAB(10)"==== (N)EW CALCULATION OR (E)ND ??? ===="
12020 U$=INKEY$
12030 IF U$="" THEN 12020
12040 IF U$="N" OR U$="n" THEN RUN
12050 GOTO 500
20000 END
```

```
90 KEY OFF
100 REM        ****************************************
110 REM        *           PROGRAM 'VAZAO3'          *
120 REM        *    WRITTEN BY DR.A./R.A. LATTERMANN  *
130 REM        *                1987                 *
140 REM        ****************************************
150 REM
160 CLS:BEEP
161 REM
162 REM UTILIZATION OF THE PROGRAM AND DATA INPUT
163 REM
170 GOSUB 2000
180 PRINT:PRINT
190 PRINT TAB(10)"THE PROGRAM 'VAZAO3' CALCULATES RUNOFF VALUES
195 PRINT TAB(10)"BY A LINEAR RESERVOIR MODEL"
200 PRINT TAB(10)"USING THE METHOD OF Z-TRANSFORMATION"
210 PRINT
220 PRINT TAB(10)"DATA OF INPUT ARE:":PRINT
240 PRINT TAB(10)"1.      TIME DISTRIBUTION OF EFFECTIVE RAIN "
250 PRINT TAB(10)"        A.) MAX. NUMBER OF TIME INTERVALS
MI":PRINT TAB(10)"        B.) EFFECTIVE RAIN VALUE (MM) /
INTERVAL":PRINT TAB(10)"            N0,N1,N2,N3,......"
260 PRINT:PRINT TAB(10)"2.      STORAGE COEFFICIENT OF THE
RESERVOIR (H)
270 PRINT:LOCATE 20,10:PRINT "        MAX. NUMBER OF TIME
INTERVALS MI = "; :INPUT MI:M=MI-1:DIM NI(M)
300 FOR I=0 TO M:LOCATE 21,17:PRINT "VALUE OF EFFECTIVE RAIN
N"; I
310 LOCATE 22,10:PRINT "
";:LOCATE 22,55 :INPUT NI(I):NEXT I:LOCATE 20,10:PRINT
SPACE$(45)
312 LOCATE 21,17:PRINT SPACE$(50):LOCATE 22,10:PRINT
SPACE$(50):PRINT:LOCATE 20,11:PRINT "         STORAGE
COEFFICIENT"; :INPUT KX:S=6:DIM HI(S):KR=1/KX
313 FOR I=0 TO S:HI(I)=KR*EXP(-KR*I):NEXT I
314 SU=0:FOR I=0 TO S:SU=SU+HI(I):NEXT I:FOR I=0 TO
S:HI(I)=HI(I)/SU:NEXT I
315 CLS:LOCATE 13,25:PRINT "W A I T    ! ! !":GOSUB 8000
390 CLS:BEEP
391 REM
392 REM DISPLAY ..
393 REM
400 GOSUB 2000
410 PRINT:PRINT
420 PRINT TAB(10)"RUNOFF CALCULATED:":PRINT
430 PRINT TAB(10)"INTERVAL (I)"; TAB(35)"RUNOFF (Q(I))"
440 FOR I=0 TO L
445 PRINT TAB(16)I; TAB(40); :PRINT USING "###.###"; Q(I):NEXT
I
450 PRINT
```

```
460 PRINT TAB(10)"==== PRINTING OF THE RESULT (Y/N) ??? ===="
470 U$=INKEY$
480 IF U$="" THEN 470
490 IF U$="Y" OR U$="y" THEN GOSUB 3000
495 GOSUB 20000:GOTO 12000
500 CLS
510 GOSUB 2000
520 PRINT:PRINT
530 PRINT TAB(10)"========= END OF THE PROGRAM ========"
540 PLAY "O4T200L8GAGAL2C":KEY ON:END
2000 PRINT TAB(10)"[ PROGRAM :    V   A   Z   A   O   3   ]
"; DATE$:PRINT:RETURN
3000 REM
3001 REM PRINTER ..
3002 REM
3009 LPRINT:LPRINT
3010 LPRINT CHR$(14); "              V   A   Z   A   O   3";
CHR$(20):LPRINT
3020 LPRINT CHR$(15); "
WRITTEN BY DR.A./R.A. LATTERMANN"; CHR$(18)
3030 LPRINT:LPRINT:LPRINT
3038 LPRINT TAB(10)"
"; DATE$:LPRINT
3040 LPRINT TAB(10)"THE PROGRAM 'VAZAO3' CALCULATES RUNOFF
VALUES "
3050 LPRINT TAB(10)"BY A LINEAR RESERVOIR MODEL USING THE
METHOD OF"
3055 LPRINT TAB(10)"Z-TRANSFORMATION ":
3060 LPRINT:LPRINT
3070 LPRINT TAB(10)"DATA OF INPUT ARE:":LPRINT
3080 LPRINT TAB(10)"1.    TIMR DISTRIBUTION OF EFFECTIVE RAIN "
3090 LPRINT TAB(10)"     A.) MAX. NUMBER OF TIME INTERVALS
MI":LPRINT TAB(10)"     B.) EFFECTIVE RAIN VALUE (MM) /
INTERVAL":LPRINT TAB(10)"         N0,N1,N2,N3, ..."
3100 LPRINT:LPRINT TAB(10)"2.    STORAGE COEFFICIENT (H)":LPRINT
3110 LPRINT TAB(10)"MAX. NUMBER OF TIME INTERVALS MI = ";
MI:LPRINT
3120 LPRINT TAB(15)"INTERVAL"; TAB(45)"EFFECTIVE RAIN VALUE
(MM)"
3140 FOR I=0 TO M
3150 LPRINT TAB(18)I; TAB(55)NI(I)
3160 NEXT I
3162 LPRINT:LPRINT TAB(10)"STORAGE COEFFICIENT"; KX:LPRINT
3170 LPRINT:LPRINT
3180 LPRINT TAB(10)"RESULTS:":LPRINT
3190 LPRINT:LPRINT TAB(10)"RUNOFF CALCULATED (MM)
[Q(I)]:":LPRINT
3200 LPRINT TAB(10)"INTERVAL:"; TAB(45)"Q(I)"
3210 FOR I=0 TO L
3220 LPRINT TAB(13)I; TAB(45)Q(I)
3230 NEXT I
3232 LPRINT:LPRINT TAB(10)"E   N   D   ..."
3240 RETURN
```

```
8000 REM
8001 REM CALCULATION ..
8002 REM
8010 DIM C(M,S)
8030 FOR I=0 TO M
8040 FOR J=0 TO S
8050 C(I,J)=HI(J)*NI(I)
8060 NEXT J
8080 NEXT I:N=0:L=M+S:DIM Q(L):FOR K=0 TO L:Q(K)=0
8090 FOR I=0 TO M
8110 FOR J=0 TO S:IF I+J=K THEN GOTO 8120
8115 GOTO 8125
8120 Q(K)=Q(K)+C(I,J)
8125 NEXT J,I,K
8130 RETURN
12000 REM
12001 REM END ..
12002 REM
12009 PRINT
12010 PRINT TAB(10)"==== (N)EW CALCULATION OR (E)ND ??? ===="
12020 U$=INKEY$
12030 IF U$="" THEN 12020
12040 IF U$="N" OR U$="n" THEN RUN
12050 GOTO 500
20000 END
```

```
90 KEY OFF
100 REM           ******************************************
110 REM           *           PROGRAM  'VAZAO4'            *
120 REM           *     WRITTEN BY DR.A./R.A. LATTERMANN   *
130 REM           *                 1987                   *
140 REM           ******************************************
150 REM
160 CLS:BEEP
161 REM
162 REM UTILIZATION OF THE PROGRAM AND DATA INPUT
163 REM
170 GOSUB 2000
180 PRINT:PRINT
190 PRINT TAB(10)"THE PROGRAM 'VAZAO4' CALCULATES RUNOFF VALUES
BY A "
200 PRINT TAB(10)"CASCADE MODEL OF TWO LINEAR RESERVOIRS USING
THE "
205 PRINT TAB(10)"METHOD OF Z-TRANSFORMATION (NASH-MODEL,N=2)"
210 PRINT
220 PRINT TAB(10)"DATA OF INPUT ARE:":PRINT
240 PRINT TAB(10)"1.      TIME DISTRIBUTION OF EFFECTIVE RAIN"
250 PRINT TAB(10)"       A.) MAX. NUMBER OF TIME INTERVALS
MI":PRINT TAB(10)"       B.) EFFECTIVE RAIN VALUE (MM) /
INTERVAL":PRINT TAB(10)"             N0,N1,N2,N3,......"
260 PRINT:PRINT TAB(10)"2.     STORAGE COEFFICIENT (H)
270 PRINT:LOCATE 20,10:PRINT "      MAX. NUMBER OF TIME
INTERVALS MI = "; :INPUT MI:M=MI-1:DIM NI(M)
300 FOR I=0 TO M:LOCATE 21,17:PRINT "EFFECTIVE RAIN VALUE N"; I
310 LOCATE 22,10:PRINT "
";:LOCATE 22,55: :INPUT NI(I):NEXT I:LOCATE 20,10:PRINT
SPACE$(45)
312 LOCATE 21,17:PRINT SPACE$(50):LOCATE 22,10:PRINT
SPACE$(50):PRINT:LOCATE 20,11:PRINT "      STORAGE
COEFFICIENT"; :INPUT KX:S=6:DIM HI(S):KR=1/KX
313 FOR I=0 TO S:HI(I)=KR^2*I*EXP(-KR*I):NEXT I
314 FOR I=0 TO S:SU=SU+HI(I):NEXT I:FOR I=0 TO
S:HI(I)=HI(I)/SU:NEXT I
315 CLS:LOCATE 13,25:PRINT "W A I T   ! ! !":GOSUB 8000
390 CLS:BEEP
391 REM
392 REM DISPLAY ..
393 REM
400 GOSUB 2000
410 PRINT:PRINT
420 PRINT TAB(10)"RUNOFF CALCULATED        :":PRINT
430 PRINT TAB(10)"INTERVAL (I)"; TAB(35)"RUNOFF (Q(I))"
440 FOR I=0 TO L
445 PRINT TAB(16)I; TAB(40); :PRINT USING "###.###"; Q(I):NEXT
I
450 PRINT
```

```
460 PRINT TAB(10)"==== PRINTING OF THE RESULT (Y/N) ??? ===="
470 U$=INKEY$
480 IF U$="" THEN 470
490 IF U$="Y" OR U$="y" THEN GOSUB 3000
495 GOSUB 20000:GOTO 12000
500 CLS
510 GOSUB 2000
520 PRINT:PRINT
530 PRINT TAB(10)"========== END OF THE PROGRAM ========"
540 PLAY "O4T200L8GAGAL2C":KEY ON:END
2000 PRINT TAB(10)"[ PROGRAM :    V   A   Z   A   O   4   ]
"; DATE$:PRINT:RETURN
3000 REM
3001 REM PRINTER ..
3002 REM
3009 LPRINT:LPRINT
3010 LPRINT CHR$(14); "               V   A   Z   A   O   4";
CHR$(20):LPRINT
3020 LPRINT CHR$(15); "
WRITTEN BY DR.A./R.A. LATTERMANN"; CHR$(18)
3030 LPRINT:LPRINT:LPRINT
3038 LPRINT TAB(10)"
"; DATE$:LPRINT
3040 LPRINT TAB(10)"THE PROGRAM 'VAZAO4' CALCULATES RUNOFF
VALUES BY A"
3050 LPRINT TAB(10)"CASCADE MODEL OF TWO LINEAR RESERVOIRS
USING THE "
3055 LPRINT TAB(10)"METHOD OF Z-TRANSFORMATION (NASH-
MODEL,N=2)"
3060 LPRINT:LPRINT
3070 LPRINT TAB(10)"DATA OF INPUT ARE:":LPRINT
3080 LPRINT TAB(10)"1.    TIME DISTRIBUTION OF EFFECTIVE RAIN"
3090 LPRINT TAB(10)"    A.) MAX. NUMBER OF TIME INTERVALS
MI":LPRINT TAB(10)"      B.) EFFECTIVE RAIN VALUE (MM) /
INTERVAL":LPRINT TAB(10)"          N0,N1,N2,N3, ..."
3100 LPRINT:LPRINT TAB(10)"2.    STORAGE COEFFICIENT (H)":LPRINT
3110 LPRINT TAB(10)"MAX. NUMBER OF TIME INTERVALS MI = ";
MI:LPRINT
3120 LPRINT TAB(15)"INTERVAL"; TAB(45)"EFFECTIVE RAIN VALUE
(MM)"
3140 FOR I=0 TO M
3150 LPRINT TAB(18)I; TAB(55)NI(I)
3160 NEXT I
3162 LPRINT:LPRINT TAB(10)"STORAGE COEFFICIENT"; KX:LPRINT
3170 LPRINT:LPRINT
3180 LPRINT TAB(10)"RESULTS:":LPRINT
3190 LPRINT:LPRINT TAB(10)"RUNOFF CALCULATED (MM)
[Q(I)]:":LPRINT
3200 LPRINT TAB(10)"INTERVAL:"; TAB(45)"Q(I)"
3210 FOR I=0 TO L
3220 LPRINT TAB(13)I; TAB(45)Q(I)
3230 NEXT I
3232 LPRINT:LPRINT TAB(10)"E   N   D ..."
```

```
j3240 RETURN
8000 REM
8001 REM CALCULATION ..
8002 REM
8010 DIM C(M,S)
8030 FOR I=0 TO M
8040 FOR J=0 TO S
8050 C(I,J)=HI(J)*NI(I)
8060 NEXT J
8080 NEXT I:N=0:L=M+S:DIM Q(L):FOR K=0 TO L:Q(K)=0
8090 FOR I=0 TO M
8110 FOR J=0 TO S:IF I+J=K THEN GOTO 8120
8115 GOTO 8125
8120 Q(K)=Q(K)+C(I,J)
8125 NEXT J,I,K
8130 RETURN
12000 REM
12001 REM END ..
12002 REM
12009 PRINT
12010 PRINT TAB(10)"==== (N)EW CALCULATION OR (E)ND ??? ===="
12020 U$=INKEY$
12030 IF U$="" THEN 12020
12040 IF U$="N." OR U$="n" THEN RUN
12050 GOTO 500
20000 END
→
```

```
90 KEY OFF
100 REM          ****************************************
110 REM          *           PROGRAM  'IDENT1'          *
120 REM          *     WRITTEN BY DR.A./R.A. LATTERMANN *
130 REM          *                 1987                 *
140 REM          ****************************************
150 REM
160 CLS:BEEP
161 REM
162 REM UTILIZATION OF THE PROGRAM AND DATA INPUT
163 REM
170 GOSUB 2000
180 PRINT:PRINT
190 PRINT TAB(10)"THE PROGRAM 'IDENT1' CALCULATES VALUES OF THE
SYSTEM "
200 PRINT TAB(10)"RESPONSE FUNCTION BY USE OF THE METHOD OF Z-
TRANSFORMATION"
210 PRINT
220 PRINT TAB(10)"DATA OF INPUT ARE:":PRINT
240 PRINT TAB(10)"1.        TIME DISTRIBUTION OF EFFECTIVE RAIN "
250 PRINT TAB(10)"       A.) MAX. NUMBER OF TIME INTERVALS
MI":PRINT TAB(10)"       B.) EFFECTIVE RAIN VALUE (MM) /
INTERVAL":PRINT TAB(10)"          N0,N1,N2,N3,......"
260 PRINT:PRINT TAB(10)"2.        CORRESPONDING RUNOFF FUNCTION "
265 PRINT TAB(10)"       A.) MAX. NUMBER OF TIME INTERVALS
SI":PRINT TAB(10)"       B.) RUNOFF (MM) / INTERVAL    ":PRINT
TAB(10)"        Q0,Q1,Q2,Q3, ..."
270 PRINT:LOCATE 20,10:PRINT "        MAX. NUMBER OF TIME
INTERVALS SI = "; :INPUT SI:DIM QI(100):S=SI-1:S1=S*5
300 FOR I=0 TO S:LOCATE 21,17:PRINT "ABFLUß Q"; I
310 LOCATE 22,10:PRINT "
";:LOCATE 22,55 :INPUT QI(I):NEXT I:LOCATE 20,10:PRINT
SPACE$(45)
311 LOCATE 20,10:PRINT SPACE$(50):LOCATE 21,10:PRINT
SPACE$(50):LOCATE 22,10:PRINT SPACE$(50)
312 PRINT:LOCATE 20,10:PRINT "        MAX. NUMBER OF TIME
INTERVALS MI = "; :INPUT MI:M=MI-1:M1=M*5:DIM NI(M1)
313 FOR I=0 TO M:LOCATE 21,17:PRINT "EFFECTIVE RAIN VALUE N"; I
314 LOCATE 22,10:PRINT "
";:LOCATE 22,55 :INPUT NI(I):NEXT I:PLAY "O2MLT250L8CEGL4O3C"
315 CLS:LOCATE 13,25:PRINT "W A I T  ! ! !":GOSUB 8000
390 CLS:BEEP
391 REM
392 REM DISPLAY ..
393 REM
400 GOSUB 2000
410 PRINT:PRINT
420 PRINT TAB(10)"CALCULATED VALUES OF THE SYSTEM RESPONSE
FUNCTION UH(I):"
:PRINT
```

```
430 PRINT TAB(10)"INTERVAL (I)"; TAB(35)"UH(I)"
440 FOR I=0 TO 10:IF ABS(HU(I))<9.999999E-05 THEN HU(I)=0
445 PRINT TAB(16)I; TAB(35); :PRINT USING "###.###"; HU(I):NEXT
I
450 PRINT
460 PRINT TAB(10)"==== PRINTING OF THE RESULT (Y/N) ??? ===="
470 U$=INKEY$
480 IF U$="" THEN 470
490 IF U$="Y" OR U$="y" THEN GOSUB 3000
495 GOSUB 20000:GOTO 12000
500 CLS
510 GOSUB 2000
520 PRINT:PRINT
530 PRINT TAB(10)"========== END OF THE PROGRAM ========"
540 PLAY "O4T200L8GAGAL2C":KEY ON:END
2000 PRINT TAB(10)"[ PROGRAM :    I  D  E  N  T  1  ]
"; DATE$:PRINT:RETURN
3000 REM
3001 REM PRINTER ..
3002 REM
3009 LPRINT:LPRINT
3010 LPRINT CHR$(14); "            I  D  E  N  T  1";
CHR$(20):LPRINT
3020 LPRINT CHR$(15); "
WRITTEN BY DR.A./R.A. LATTERMANN"; CHR$(18)
3030 LPRINT:LPRINT:LPRINT
3038 LPRINT TAB(10)"
"; DATE$:LPRINT
3040 LPRINT TAB(10)"THE PROGRAM 'IDENT1' CALCULATES VALUES OF
THE SYSTEM "
3050 LPRINT TAB(10)"RESPONSE FUNCTION BY USE OF THE METHOD OF
Z-TRANSFORMATION"
3060 LPRINT:LPRINT
3070 LPRINT TAB(10)"DATA OF INPUT ARE:":LPRINT
3080 LPRINT TAB(10)"1.    TIME DISTRIBUTION OF EFFECTIVE RAIN"
3090 LPRINT TAB(10)"     A.) MAX. NUMBER OF TIME INTERVALS
MI":LPRINT TAB(10)"     B.) EFFECTIVE RAIN VALUE (MM) /
INTERVAL":LPRINT TAB(10)"         N0,N1,N2,N3, ..."
3100 LPRINT:LPRINT TAB(10)"2.    CORRESPONDING RUNOFF FUNCTION"
3105 LPRINT TAB(10)"     A.) MAX. NUMBER OF TIME INTERVALS
SI":LPRINT TAB(10)"     B.) RUNOFF (MM) / INTERVAL ":LPRINT
TAB(10)"         Q0,Q1,Q2,Q3, ...":LPRINT
3110 LPRINT TAB(10)"MAX. NUMBER OF TIME INTERVALS MI = ";
MI:LPRINT
3120 LPRINT TAB(15)"INTERVAL"; TAB(45)"EFFECTIVE RAIN VALUE
(MM)"
3140 FOR I=0 TO M
3150 LPRINT TAB(18)I; TAB(55)NI(I)
3160 NEXT I
3162 LPRINT:LPRINT TAB(10)"MAX. NUMBER OF TIME INTERVALS SI =
"; SI:LPRINT
3164 LPRINT TAB(15)"INTERVAL"; TAB(45)"RUNOFF (MM)"
3166 FOR I=0 TO S
```

```
3168 LPRINT TAB(18)I; TAB(50)QI(I)
3169 NEXT I
3170 LPRINT:LPRINT:LPRINT
3180 LPRINT TAB(10)"RESULTS:":LPRINT
3190 LPRINT:LPRINT TAB(10)"CALCULATED VALUES OF THE SYSTEM
RESPONSE FUNCTION UH(I) :":LPRINT
3200 LPRINT TAB(10)"INTERVAL:"; TAB(45)"UH(I)"
3210 FOR I=0 TO 10:IF ABS(HU(I))<9.999999E-05 THEN HU(I)=0
3220 LPRINT TAB(13)I; TAB(45)HU(I)
3230 NEXT I
3232 LPRINT:LPRINT TAB(10)"E   N   D   ..."
3240 RETURN
8000 REM
8001 REM CALCULATION ..
8002 REM
8010 DIM HU(20),K(20)
8020 HU(0)=(1/NI(0))*QI(0)
8030 HU(1)=(1/NI(0))*(QI(1)-HU(0)*NI(1))
8040 FOR I=2 TO 10
8050 K(0)=HU(0)*NI(I)
8060 FOR J=1 TO I-1
8070 K(J)=K(J-1)+HU(J)*NI(I-J)
8080 NEXT J
8090 HU(I)=(1/NI(0))*(QI(I)-K(I-1))
8100 NEXT I
8110 RETURN
12000 REM
12001 REM END ..
12002 REM
12009 PRINT
12010 PRINT TAB(10)"==== (N)EW CALCULATION OR (E)ND ??? ===="
12020 U$=INKEY$
12030 IF U$="" THEN 12020
12040 IF U$="N" OR U$="n" THEN RUN
12050 GOTO 500
20000 END
```

```
90 KEY OFF
100 REM          ****************************************
110 REM          *          PROGRAM  'IDENT2'          *
120 REM          *    WRITTEN BY DR.A./R.A. LATTERMANN  *
130 REM          *               1987                 *
140 REM          ****************************************
150 REM
160 CLS:BEEP
161 REM
162 REM UTILIZATION OF THE PROGRAM AND DATA INPUT
163 REM
170 GOSUB 2000
180 PRINT:PRINT
190 PRINT TAB(10)"THE PROGRAM 'IDENT2' CALCULATES VALUES OF THE
SYSTEM"
200 PRINT TAB(10)"RESPONSE FUNCTION BY USE OF THE  MATRIX
METHOD"
210 PRINT
220 PRINT TAB(10)"DATA OF INPUT ARE:":PRINT
240 PRINT TAB(10)"1.     TIME DISTRIBUTION OF EFFECTIVE RAIN"
250 PRINT TAB(10)"      A.) MAX. NUMBER OF TIME INTERVALS
MI":PRINT TAB(10)"      B.) EFFECTIVE RAIN VALUE (MM) /
INTERVAL":PRINT TAB(10)"          N0,N1,N2,N3,......"
260 PRINT:PRINT TAB(10)"2.     CORRESPONDING RUNOFF FUNCTION"
265 PRINT TAB(10)"      A.) MAX. NUMBER OF TIME INTERVALS
K":PRINT TAB(10)"      B.) RUNOFF (MM) / INTERVAL    ":PRINT
TAB(10)"          Q0,Q1,Q2,Q3, ..."
270 PRINT:LOCATE 20,10:PRINT "       MAX. NUMBER OF TIME
INTERVALS K = "; :INPUT K:DIM Q(K)
300 FOR I=1 TO K:LOCATE 21,17:PRINT "RUNOFF VALUE Q :"; I-1
310 LOCATE 22,10:PRINT "
";:LOCATE 22,55  :INPUT Q(I):NEXT I:LOCATE 20,10:PRINT
SPACE$(45)
311 LOCATE 21,17:PRINT SPACE$(50):LOCATE 20,17:PRINT
SPACE$(50):LOCATE 22,10:PRINT SPACE$(60)
312 PRINT:LOCATE 20,10:PRINT "       MAX. NUMBER OF TIME
INTERVALS M = "; :INPUT M:R=K-M+1:DIM NI(K,R)
313 FOR I=1 TO M:LOCATE 21,17:PRINT "EFFECTIVE RAIN VALUE N:";
I-1
314 LOCATE 22,10:PRINT "
";:LOCATE 22,55  :INPUT NI(I,1):NEXT I:PLAY
"O2MLT250L8CEGL4O3C"
315 CLS:LOCATE 13,25:PRINT "W A I T  ! ! !":GOSUB 8000
390 CLS:BEEP
391 REM
392 REM DISPLAY ..
393 REM
400 GOSUB 2000
410 PRINT:PRINT
```

```
420 PRINT TAB(10)"CALCULATED SYSTEM RESPONSE FUNCTION UH(I)
:":PRINT
430 PRINT TAB(10)"INTERVAL (I)"; TAB(35)"UH(I)"
440 FOR I=1 TO R:U(I)=INT(U(I)*100+.5)/100
445 PRINT TAB(16)I-1; TAB(35); :PRINT USING "###.###";
U(I):NEXT I
450 PRINT
460 PRINT TAB(10)"==== PRINTING OF THE RESULT (Y/N) ??? ===="
470 U$=INKEY$
480 IF U$="" THEN 470
490 IF U$="Y" OR U$="y" THEN GOSUB 3000
495 GOSUB 20000:GOTO 12000
500 CLS
510 GOSUB 2000
520 PRINT:PRINT
530 PRINT TAB(10)"========== END OF THE PROGRAM ========="
540 PLAY "O4T200L8GAGAL2C":KEY ON:END
2000 PRINT TAB(10)"[ PROGRAM :   I   D   E   N   T   2   ]
"; DATE$:PRINT:RETURN
3000 REM     .
3001 REM PRINTER ..
3002 REM
3009 LPRINT:LPRINT
3010 LPRINT CHR$(14); "            I   D   E   N   T   2";
CHR$(20):LPRINT
3020 LPRINT CHR$(15); "
WRITTEN BY DR.A./R.A. LATTERMANN"; CHR$(18)
3030 LPRINT:LPRINT:LPRINT
3038 LPRINT TAB(10)"
"; DATE$:LPRINT
3040 LPRINT TAB(10)"THE PROGRAM 'IDENT2' CALCULATES VALUES OF
THE SYSTEM"
3050 LPRINT TAB(10)"RESPONSE FUNCTION BY USE OF THE MATRIX
METHOD"
3060 LPRINT:LPRINT
3070 LPRINT TAB(10)"DATA OF INPUT ARE:":LPRINT
3080 LPRINT TAB(10)"1.    TIME DISTRIBUTION OF EFFECTIVE RAIN"
3090 LPRINT TAB(10)"     A.) MAX. NUMBER OF TIME INTERVALS
MI":LPRINT TAB(10)"     B.) EFFECTIVE RAIN VALUE (MM) /
INTERVAL":LPRINT TAB(10)"         N0,N1,N2,N3, ..."
3100 LPRINT:LPRINT TAB(10)"2.    CORRESPONDING RUNOFF FUNCTION"
3105 LPRINT TAB(10)"     A.) MAX. NUMBER OF TIME INTERVALS
K":LPRINT TAB(10)"     B.) RUNOFF (MM) / INTERVAL ":LPRINT
TAB(10)"         Q0,Q1,Q2,Q3, ...":LPRINT
3110 LPRINT TAB(10)"MAX. NUMBER OF TIME INTERVALS M = ";
M:LPRINT
3120 LPRINT TAB(15)"INTERVAL"; TAB(45)"EFFECTIVE RAIN VALUE
(MM)"
3140 FOR I=1 TO M
3150 LPRINT TAB(18)I-1; TAB(55)NI(I,1)
3160 NEXT I
3162 LPRINT:LPRINT TAB(10)"MAX. NUMBER OF TIME INTERVALS K = ";
K:LPRINT
```

```
3164 LPRINT TAB(15)"INTERVAL"; TAB(45)"RUNOFF (MM)"
3166 FOR I=1 TO K
3168 LPRINT TAB(18)I-1; TAB(50)Q(I)
3169 NEXT I
3170 LPRINT:LPRINT:LPRINT
3180 LPRINT TAB(10)"RESULTS :":LPRINT
3190 LPRINT:LPRINT TAB(10)"CALCULATED VALUES OF THE SYSTEM
RESPONSE FUNCTION UH(I):" :LPRINT
3200 LPRINT TAB(10)"INTERVAL:"; TAB(45)"UH(I)"
3210 FOR I=1 TO R:U(I)=INT(U(I)*100+.5)/100
3220 LPRINT TAB(13)I; TAB(45)U(I)
3230 NEXT I
3232 LPRINT:LPRINT TAB(10)"E   N   D ..."
3240 RETURN
8000 REM
8001 REM CALCULATION ..
8002 REM
8010 T=1
8020 FOR I=2 TO R
8030 FOR J=1 TO K-T
8040 NI(J+T,I)=NI(J,1)
8050 NEXT J
8060 T=T+1
8070 NEXT I
8080 DIM NT(R,K)
8090 FOR I=1 TO K
8100 FOR J=1 TO R
8110 NT(J,I)=NI(I,J)
8120 NEXT J
8130 NEXT I
8140 DIM NM(R,R)
8150 SU=0
8160 FOR L=1 TO R
8170 FOR I=1 TO R
8180 FOR J=1 TO K
8190 SU=SU+(NT(L,J))*(NI(J,I))
8200 NEXT J
8210 NM(L,I)=SU
8220 SU=0
8230 NEXT I
8240 NEXT L
8250 DIM E(R,R)
8260 FOR I=1 TO R
8270 FOR J=1 TO R
8280 IF J=I THEN 8310
8290 E(I,J)=0
8300 GOTO 8320
8310 E(I,J)=1
8320 NEXT J,I
8330 DIM NN(R,R),EN(R,R)
8340 T=1
8350 FOR L=1 TO R-1
8360 T=T+1
```

```
8370 FOR I=T TO R
8380 FOR J=1 TO R
8390 NN(I,J)=NM(I,J)-(NM(I,L)/NM(L,L))*NM(L,J)
8400 EN(I,J)=E(I,J)-(NM(I,L)/NM(L,L))*E(L,J)
8410 NEXT J
8420 NEXT I
8430 FOR I=T TO R
8440 FOR J=1 TO R
8450 NM(I,J)=NN(I,J)
8460 E(I,J)=EN(I,J)
8470 NEXT J
8480 NEXT I
8485 NEXT L
8490 FOR I=1 TO R
8500 NN(1,I)=NM(1,I)
8510 EN(1,I)=E(1,I)
8520 NEXT I
8530 DIM A(R,R)
8540 V=0
8550 FOR I=1. TO R
8560 FOR J=0 TO R-1
8570 FOR C=1 TO R
8580 V=V+NN(R-J,C)*A(C,I)
8590 NEXT C
8600 VN=V+NN(R-J,R-J)*A(R-J,I)
8610 A(R-J,I)=(EN(R-J,I)-VN)/NN(R-J,R-J)
8620 VN=0:V=0
8630 NEXT J
8640 NEXT I
8650 DIM B(R,K)
8660 KU=0
8670 FOR L=1 TO R
8680 FOR I=1 TO K
8690 FOR J=1 TO R
8700 KU=KU+A(L,J)*NT(J,I)
8710 NEXT J
8720 B(L,I)=KU
8730 KU=0
8740 NEXT I
8750 NEXT L
8760 DIM U(R)
8770 CU=0
8780 FOR I=1 TO R
8790 FOR J=1 TO K
8800 CU=CU+B(I,J)*Q(J)
8810 NEXT J
8820 U(I)=CU
8830 CU=0
8840 NEXT I
8850 RETURN
12000 REM
12001 REM END ..
12002 REM
```

```
12009 PRINT
12010 PRINT TAB(10)"==== (N)EW CALCULATION OR (E)ND ??? ===="
12020 U$=INKEY$
12030 IF U$="" THEN 12020
12040 IF U$="N" OR U$="n" THEN RUN
12050 GOTO 500
20000 END
```

```
10 REM
20 REM
*****************************************************************
******
30 REM *          PROGRAM :          P     C     O     R     R
..        *
40 REM *
===========================================================     *
50 REM *
*
60 REM *             WRITTEN BY DR.A./R.A. LATTERMANN
*
70 REM *
*
80 REM *                          1987
*
90 REM *
*
100 REM
*****************************************************************
******
110 REM
120 CLS
130 KEY OFF
131 REM
132 REM UTILIZATION OF THE PROGRAM AND DATA INPUT
133 REM
140 BEEP
150 GOSUB 2000
160 PRINT
170 PRINT TAB(10)"THE PROGRAM 'PCORR' CALCULATES THE
CORRELATION AND"
180 PRINT TAB(10)"AUTOCORRELATION FUNCTION OF TWO STOCHASTIC
PROCESSES:"
190 PRINT TAB(10)"( PRECIPITATION AND RUNOFF, DIMENSION: MM)."
200 PRINT
210 PRINT
220 LOCATE 12,10:PRINT "MAX. NUMBER OF EVENTS"; :INPUT N1
230 LOCATE 12,10:PRINT SPACE$(50)
240 LOCATE 12,10:PRINT "THE STOCHASTIC PROCESS OF
PRECIPITATION:"
250 LOCATE 14,10:PRINT "MAX. NUMBER OF TIME INTERVALS/EVENT";
:INPUT N2
260 DIM P(N1,N2),SP(N2),SPP(N2,N2),CPP(N2,N2),SMP(N2),AMP(N2)
270 FOR I=1 TO N1
280 FOR J=1 TO N2
290 LOCATE 15,10:PRINT "VALUE"; J; "OF THE EVENT"; I; "OF
PRECIPITATION"; :INPUT P(I,J)
300 LOCATE 15,10:PRINT SPACE$(50)
310 NEXT J,I
```

```
320 LOCATE 12,10:PRINT SPACE$(50):LOCATE 14,10:PRINT SPACE$(50)
330 LOCATE 12,10:PRINT "THE STOCHASTIC PROCESS OF RUNOFF:"
340 LOCATE 14,10:PRINT "MAX NUMBER OF TIME INTERVALS/EVENT";
:INPUT N3
350 A=N2
360 IF N3<N2 THEN A=N3
370 DIM
Q(N1,N3),SQ(N3),SQQ(N3,N3),SQP(A,A),CQQ(N3,N3),CQP(A,A),SMQ(N3)
,AMQ(N3),CMP(A),MPQ(A)
380 FOR I=1 TO N1
390 FOR J=1 TO N3
400 LOCATE 15,10:PRINT "VALUE"; J; "OF THE EVENT"; I; " OF
RUNOFF:"; :INPUT Q(I,J)
410 LOCATE 15,10:PRINT SPACE$(50)
420 NEXT J,I
422 REM
425 REM DISPLAY ..
428 REM
430 GOSUB 5000:GOSUB 8000
440 CLS
450 GOSUB 2000
460 PRINT
470 PRINT TAB(10)"AUTOCORRELATION MATRIX NORMALIZED OF
PRECIPITATION AND"
480 PRINT TAB(10)"                    AVERAGES"
490 PRINT
500 FOR I=1 TO N2
510 FOR J=1 TO N2
520 UK=ACP(I,J):RP=INT(UK*100+.5)/100
530 PRINT "RP("; I; ","; J; ") = "; RP,
540 NEXT J,I
550 PRINT
560 PRINT TAB(10)"===> PLEASE PRESS ANY KEY..."
570 U$=INKEY$
580 IF U$="" THEN 570
590 CLS
600 GOSUB 2000:PRINT
610 PRINT TAB(10)".. CONTINUE .."
620 PRINT
640 FOR I=1 TO N2
650 AMP=INT(AMP(I)*100+.5)/100
660 PRINT "MRPP("; I; ") = "; AMP,
670 NEXT I
680 PRINT:GOSUB 690:GOTO 720
690 PRINT TAB(10)"===> PLEASE PRESS ANY KEY (CTRL+C ==> STOP)
..."
700 U$=INKEY$
710 IF U$="" THEN 700
715 RETURN
720 CLS
730 GOSUB 2000:PRINT
740 PRINT TAB(10)"AUTOCORRELATION MATRIX NORMALIZED OF RUNOFF
AND"
```

```
750 PRINT TAB(10)"                    AVERAGES"
760 PRINT:PRINT
770 FOR I=1 TO N3
780 FOR J=1 TO N3
790 RQ=INT(ACQ(I,J)*100+.5)/100
800 PRINT "RQ("; I; ","; J; ") = "; RQ,
810 NEXT J,I
815 PRINT
820 FOR I=1 TO N3
830 AMQ=INT(AMQ(I)*100+.5)/100
840 PRINT "MRQQ("; I; ") = "; AMQ,
850 NEXT I
860 PRINT:GOSUB 690
870 CLS
880 BEEP
890 GOSUB 2000:PRINT
900 PRINT TAB(10)"CORRELATION MATRIX NORMALIZED"
910 PRINT TAB(10)"     AND AVERAGES"
920 PRINT
940 FOR I=1 TO A
950 FOR J=1 TO A
960 RA=INT(COR(I,J)*100+.5)/100
970 PRINT "RPQ("; I; ","; J; ") = "; RA,
980 NEXT J,I
990 PRINT
995 PRINT
1000 FOR I=1 TO A
1010 CMP=INT(CMP(I)*100+.5)/100
1020 PRINT "MRPQ("; I; ") = "; CMP,
1030 NEXT I
1040 PRINT:GOSUB 690:PRINT
1050 CLS:GOSUB 2000:PRINT
1060 PRINT TAB(10)"AUTOCORRELATION MATRIX OF PRECIPITATION"
1070 PRINT TAB(10)"               AND AVERAGES"
1080 PRINT
1100 FOR I=1 TO N2
1110 FOR J=1 TO N2
1120 R1=INT((SPP(I,J)/N1)*100+.5)/100
1130 PRINT "CPP("; I; ","; J; ") = "; R1,
1140 NEXT J,I
1150 PRINT
1155 PRINT
1160 FOR I=1 TO N2
1170 SMP=INT(SMP(I)*100+.5)/100
1180 PRINT   "MCPP("; I; ") = "; SMP,
1190 NEXT I
1200 PRINT:GOSUB 690:PRINT
1210 CLS:PRINT:GOSUB 2000:PRINT
1220 PRINT TAB(10)"AUTOCORRELATION MATRIX OF RUNOFF "
1230 PRINT TAB(10)"                 AND AVERAGES"
1240 PRINT
1250 FOR I=1 TO N3
1260 FOR J=1 TO N3
```

```
1270 R2=INT((SQQ(I,J)/N1)*100+.5)/100
1280 PRINT "CQQ("; I; ","; J; ") = "; R2,
1290 NEXT J,I
1300 PRINT
1305 PRINT
1310 FOR I=1 TO N3
1320 SMQ=INT(SMQ(I)*100+.5)/100
1330 PRINT "MCQQ("; I; ") = "; SMQ,
1340 NEXT I
1350 PRINT:GOSUB 690:PRINT
1360 CLS:PRINT:GOSUB 2000:PRINT
1370 PRINT TAB(10)"CORRELATION MATRIX AND AVERAGES "
1390 PRINT
1400 FOR I=1 TO A
1410 FOR J=1 TO A
1420 R3=INT((SQP(I,J)/N1)*100+.5)/100
1430 PRINT "CQP("; I; ","; J; ") = "; R3,
1440 NEXT J,I
1450 PRINT
1460 FOR I=1 TO A
1470 MPQ=INT(MPQ(I)*100+.5)/100
1480 PRINT "MCQP("; I; ") = "; MPQ,
1490 NEXT I
1500 PRINT
1510 PRINT
1520 PRINT TAB(10)"===> PRINTING OF THE RESULT (Y/N) ???"
1530 U$=INKEY$
1540 IF U$="" THEN 1530
1550 IF U$="Y" OR U$="y" THEN GOSUB 11000:GOTO 1600
1560 GOTO 1600
1600 CLS
1610 PRINT:GOSUB 2000:PRINT
1620 PRINT TAB(10)"===== END OF THE PROGRAM 'PCORR' ..."
1630 PRINT:PRINT:PRINT:PRINT:PRINT
1640 KEY ON
1650 END
2000 PRINT
2010 PRINT TAB(10)"PROGRAM: [ P   C   O   R   R ]
"; DATE$
2020 RETURN
5000 CLS
5010 LOCATE 10,20:PRINT "=====> W A I T <====="
5020 RETURN
8000 REM
8010 REM CALCULATION
8020 REM
8030 FOR I=1 TO N2
8040 FOR J=1 TO N1
8050 SP(I)=SP(I)+P(J,I)
8060 NEXT J,I
8070 FOR I=1 TO N3
8080 FOR J=1 TO N1
8090 SQ(I)=SQ(I)+Q(J,I)
```

```
8100 NEXT J,I
8110 FOR I=1 TO N2
8120 FOR K=0 TO N2-I
8130 FOR J=1 TO N1
8140 SPP(I,I+K)=SPP(I,I+K)+P(J,I+K)*P(J,I)
8150 CPP(I,I+K)=CPP(I,I+K)+(P(J,I+K)-SP(I+K)/N1)*(P(J,I)-
SP(I)/N1)
8160 NEXT J,K,I
8170 FOR I=1 TO N3
8180 FOR K=0 TO N3-I
8190 FOR J=1 TO N1
8200 SQQ(I,I+K)=SQQ(I,I+K)+Q(J,I+K)*Q(J,I)
8210 CQQ(I,I+K)=CQQ(I,I+K)+(Q(J,I+K)-SQ(I+K)/N1)*(Q(J,I)-
SQ(I)/N1)
8220 NEXT J,K,I
8230 FOR I=1 TO A
8240 FOR K=0 TO A-I
8250 FOR J=1 TO N1
8260 SQP(I,I+K)=SQP(I,I+K)+Q(J,I+K)*P(J,I)
8270 CQP(I,I+K)=CQP(I,I+K)+(Q(J,I+K)-SQ(I+K)/N1)*(P(J,I)-
SP(I)/N1)
8280 NEXT J,K,I
8290 FOR I=1 TO N2
8300 FOR J=1 TO N2
8310 SPP(J,I)=SPP(I,J)
8320 CPP(J,I)=CPP(I,J)
8330 ACP(I,J)=CPP(I,J)/SQR(CPP(J,J)*CPP(I,I))
8340 NEXT J,I
8350 FOR I=0 TO N2-1
8360 FOR J=1 TO N2-I
8370 AMP(I+1)=AMP(I+1)+ACP(J+I,J)
8380 SMP(I+1)=SMP(I+1)+SPP(J+I,J)
8390 NEXT J
8400 AMP(I+1)=AMP(I+1)/(N2-I)
8410 SMP(I+1)=SMP(I+1)/(N2-I)
8420 NEXT I
8430 FOR I=1 TO N3
8440 FOR J=1 TO N3
8450 SQQ(J,I)=SQQ(I,J)
8460 CQQ(J,I)=CQQ(I,J)
8470 ACQ(I,J)=CQQ(I,J)/SQR(CQQ(J,J)*CQQ(I,I))
8480 NEXT J,I
8490 FOR I=0 TO N3-1
8500 FOR J=1 TO N3-I
8510 AMQ(I+1)=AMQ(I+1)+ACQ(J+I,J)
8520 SMQ(I+1)=SMQ(I+1)+SQQ(J+I,J)
8530 NEXT J
8540 AMQ(I+1)=AMQ(I+1)/(N3-I)
8550 SMQ(I+1)=SMQ(I+1)/(N3-I)
8560 NEXT I
8570 FOR I=1 TO A
8580 FOR J=1 TO A
8590 SQP(J,I)=SQP(I,J)
```

```
8600 CQP(J,I)=CQP(I,J)
8610 COR(I,J)=CQP(I,J)/SQR(CQQ(J,J)*CPP(I,I))
8620 NEXT J,I
8630 FOR I=0 TO A-1
8640 FOR J=1 TO A-I
8650 CMP(I+1)=CMP(I+1)+COR(J+I,J)
8660 MPQ(I+1)=MPQ(I+1)+SQP(J+I,J)
8670 NEXT J
8680 CMP(I+1)=CMP(I+1)/(A-I)
8690 MPQ(I+1)=MPQ(I+1)/(A-I)
8700 NEXT I
8710 RETURN
11000 REM
11010 REM PRINTER ..
11020 REM
11030 LPRINT:LPRINT
11040 LPRINT CHR$(14); "              P   C   O   R   R";
CHR$(20):LPRINT
11050 LPRINT CHR$(15); "
WRITTEN BY R.A./R.A. LATTERMANN"; CHR$(18)
11070 LPRINT:LPRINT:LPRINT
11090 LPRINT TAB(10)"
"; DATE$:LPRINT
11170 LPRINT TAB(10)"THE PROGRAM 'PCORR' CALCULATES CORRELATION
AND AUTO CORRELATION-"
11180 LPRINT TAB(10)"FUNCTION OF TWO STOCHASTIC PROCESSES:"
11190 LPRINT TAB(10)"(PRECIPITATION AND RUNOFF, DIMENSION:
MM)."
11200 LPRINT
11210 LPRINT
11220 LPRINT TAB(10)"MAX. NUMBER OF EVENTS"; N1
11230 LPRINT
11240 LPRINT TAB(10)"THE STOCHASTIC PROCESS OF PRECIPITATION:"
11250 LPRINT TAB(10)"MAX. NUMBER OF TIME INTERVALS/EVENT"; N2
11255 LPRINT
11270 FOR I=1 TO N1
11280 FOR J=1 TO N2
11290 LPRINT TAB(10)"VALUE"; J; " OF THE EVENT "; I; " OF
PRECIPITATION:"; P(I,J)
11310 NEXT J,I:LPRINT
11330 LPRINT TAB(10)"THE STOCHASTIC PROCESS OF RUNOFF:"
11340 LPRINT TAB(10)"MAX. NUMBER OF TIME INTERVALS/EVENT"; N3
11350 LPRINT
11380 FOR I=1 TO N1
11390 FOR J=1 TO N3
11400 LPRINT TAB(10)"VALUE"; J; " OF THE EVENT "; I; " OF
RUNOFF:";   Q(I,J)
11420 NEXT J,I:LPRINT CHR$(12)
11430 LPRINT TAB(10)"RESULTS:"
11440 LPRINT TAB(10)"==========":LPRINT
11470 LPRINT TAB(10)"AUTOCORRELATION FUNCTION NORMALIZED OF
PRECIPITATION"
11480 LPRINT TAB(10)"                    AND AVERAGES"
```

```
11490 LPRINT
11500 FOR I=1 TO N2
11510 FOR J=1 TO N2
11520 UK=ACP(I,J):RP=INT(UK*100+.5)/100
11530 LPRINT "   RP("; I; ","; J; ") = "; :LPRINT USING
"###.###"; RP; :LPRINT"             ";
11540 NEXT J,I
11550 LPRINT
11640 FOR I=1 TO N2
11650 AMP=INT(AMP(I)*100+.5)/100
11660 LPRINT "   MRPP("; I; ") = "; :LPRINT USING "###.###";
AMP; :LPRINT"             ";
11670 NEXT I:LPRINT
11675 LPRINT
11740 LPRINT TAB(10)"AUTOCORRELATION FUNCTION NORMALIZED OF
RUNOFF "
11750 LPRINT TAB(10)"              AND AVERAGES"
11760 LPRINT
11770 FOR I=1 TO N3
11780 FOR J=1 TO N3
11790 RQ=INT(ACQ(I,J)*100+.5)/100
11800 LPRINT "   RQQ("; I; ","; J; ") = "; :LPRINT USING
"###.###"; RQ; :LPRINT"             ";
11810 NEXT J,I:LPRINT
11820 FOR I=1 TO N3
11830 AMQ=INT(AMQ(I)*100+.5)/100
11840 LPRINT "   MRQQ("; I; ") = "; :LPRINT USING "###.###";
AMQ; :LPRINT"             ";
11850 NEXT I
11895 LPRINT:LPRINT
11900 LPRINT TAB(10)"CORRELATION MATRIX NORMALIZED"
11910 LPRINT TAB(10)"     AND AVERAGES"
11920 LPRINT
11940 FOR I=1 TO A
11950 FOR J=1 TO A
11960 RA=INT(COR(I,J)*100+.5)/100
11970 LPRINT "   RPQ("; I; ","; J; ") = "; :LPRINT USING
"###.###"; RA; :LPRINT"             ";
11980 NEXT J,I
11990 LPRINT
12000 FOR I=1 TO A
12010 CMP=INT(CMP(I)*100+.5)/100
12020 LPRINT "   MRPQ("; I; ") = "; :LPRINT USING "###.###";
CMP; :LPRINT"             ";
12030 NEXT I:LPRINT
12035 LPRINT
12060 LPRINT TAB(10)"AUTOCORRELATION FUNCTION OF PRECIPITATION"
12070 LPRINT TAB(10)"              AND AVERAGES"
12080 LPRINT
12100 FOR I=1 TO N2
12110 FOR J=1 TO N2
12120 R1=INT((SPP(I,J)/N1)*100+.5)/100
```

```
12130 LPRINT "   CPP("; I; ","; J; ") = "; :LPRINT USING
"###.###"; R1; :LPRINT"                    ";
12140 NEXT J,I
12150 LPRINT
12160 FOR I=1 TO N2
12170 SMP=INT(SMP(I)*100+.5)/100
12180 LPRINT "   MCPP("; I; ") = "; :LPRINT USING "###.###";
SMP; :LPRINT"                    ";
12190 NEXT I:LPRINT
12195 LPRINT
12220 LPRINT TAB(10)"AUTOCORRELATION FUNCTION OF RUNOFF "
12230 LPRINT TAB(10)"                AND AVERAGES"
12240 LPRINT
12250 FOR I=1 TO N3
12260 FOR J=1 TO N3
12270 R2=INT((SQQ(I,J)/N1)*100+.5)/100
12280 LPRINT "   CQQ("; I; ","; J; ") = "; :LPRINT USING
"###.###"; R2; :LPRINT"                    ";
12290 NEXT J,I
12300 LPRINT
12310 FOR I=1 TO N3
12320 SMQ=INT(SMQ(I)*100+.5)/100
12330 LPRINT "   MCQQ("; I; ") = "; :LPRINT USING "###.###";
SMQ; :LPRINT"                    ";
12340 NEXT I:LPRINT
12350 LPRINT
12370 LPRINT TAB(10)"CORRELATION FUNCTION AND AVERAGES"
12390 LPRINT
12400 FOR I=1 TO A
2410 FOR J=1 TO A
12420 R3=INT((SQP(I,J)/N1)*100+.5)/100
12430 LPRINT "   CQP("; I; ","; J; ") = "; :LPRINT USING
"###.###"; R3; :LPRINT"                    ";
12440 NEXT J,I
12450 LPRINT
12460 FOR I=1 TO A
12470 MPQ=INT(MPQ(I)*100+.5)/100
12480 LPRINT "   MCQP("; I; ") = "; :LPRINT USING "###.###";
MPQ; :LPRINT"                    ";
12490 NEXT I.
12495 LPRINT:LPRINT
12500 LPRINT:LPRINT TAB(10)" .. E N D .."
12510 LPRINT CHR$(12):RETURN
→
```

Subject Index